ALSO BY GEORGE DYSON

Turing's Cathedral

Project Orion

Darwin Among the Machines

Baidarka

ANALOGIA

FARRAR, STRAUS AND GIROUX

NEW YORK

ANALOGIA

THE EMERGENCE
OF TECHNOLOGY
BEYOND
PROGRAMMABLE
CONTROL

GEORGE DYSON

View *of the* Land *in* PRINCE WILLIAM'S Sound, *taken from the* first Anchoring
to the Northward *of* Cape Hinchingbrook

Farrar, Straus and Giroux
120 Broadway, New York 10271

Printed in the United States of America
First edition, 2020

Library of Congress Cataloging-in-Publication Data
Names: Dyson, George, 1953– author.
Title: Analogia : the emergence of technology beyond programmable control / George Dyson.
Description: First edition. | New York : Farrar, Straus and Giroux, 2020. | Includes
  bibliographical references and index.
Identifiers: LCCN 2020012300 | ISBN 9780374104863 (hardback)
Subjects: LCSH: Technology and state—United States—History. | Science and state—United
  States—History. | Technology—Forecasting. | Electronic digital computers—Philosophy.
Classification: LCC T21 .D97 2020 | DDC 303.48/30973—dc23
LC record available at https://lccn.loc.gov/2020012300

Designed by Gretchen Achilles

Our books may be purchased in bulk for promotional, educational,
or business use. Please contact your local bookseller or the Macmillan Corporate
and Premium Sales Department at 1-800-221-7945, extension 5442,
or by e-mail at MacmillanSpecialMarkets@macmillan.com.

www.fsgbooks.com
www.twitter.com/fsgbooks • www.facebook.com/fsgbooks

1   3   5   7   9   10   8   6   4   2

Once I moved about like the wind.

—GOYAAŁÉ (GERONIMO), 1886

CONTENTS

0.

THE LEIBNIZ ARCHIPELAGO

3

1.

1741

13

2.

LAST OF THE APACHES

39

3.

AGE OF REPTILES

71

4.

VOICE OF THE DOLPHINS

105

5.

TREE HOUSE

137

6.

STRING THEORY

167

7.

EREWHON REVISITED

199

8.

NO TIME IS THERE

221

9.

CONTINUUM HYPOTHESIS

237

Notes   257

Acknowledgments   277

Index   279

ANALOGIA

0.

THE LEIBNIZ ARCHIPELAGO

From Analog to Digital and Back

In July 1716, Gottfried Wilhelm Leibniz, a seventy-year-old lawyer, philosopher, and mathematician whose "tragedy was that he met the lawyers before the scientists," joined Peter the Great, the forty-four-year-old tsar of Russia, in taking the cure at Bad Pyrmont in Saxony, drinking mineral water instead of alcohol for the duration of their eight-day stay.[1]

Leibniz, who would be dead within the year, laid three grand projects before the tsar. First was a proposal to send an overland expedition across Siberia to the Kamchatka Peninsula and the Pacific, where one or more oceangoing vessels would be launched on a voyage of discovery to determine whether Asia and America were separated, and if so, where? What languages were spoken by the inhabitants, and could this shed light on the origins and evolution of the human race? Were the rivers navigable? How does the magnetic declination vary with location, and does it also vary in time? What lay between the Russian far east and the American Northwest? Could Russia extend its claims?

View of the Land on the Coast of AMERICA when the Volcano bore N.NE. 2 E. and the Sou. part of Oonemak N.40 W. dis 7 lea.[2]

Second was a proposal to establish a Russian academy of sciences, modeled on the success of the existing European academies while leaving their infirmities behind.

Third was a plan to use digital computers "to work out, by an infallible calculus, the doctrines most useful for life, that is, those of morality and metaphysics," by encoding natural language and its underlying concepts through a numerical mapping to an alphabet of primes.[2] Leibniz sought Peter's support to introduce this *calculus ratiocinator* to China, whose philosophers he credited with the invention of binary arithmetic, and to adopt this system in the tsar's campaign for the modernization and expansion of Russia, which Leibniz saw as a tabula rasa, or blank slate, upon which his vision of a rational society based on science, logic, and machine intelligence might play out.

"The human race will have a new kind of instrument which will increase the power of the mind much more than optical lenses strengthen the eyes," he argued. "Reason will be right beyond all doubt only when it is everywhere as clear and certain as only arithmetic has been until now."[3] Leibniz observed that the functions of binary arithmetic correspond to the logical operations of "and," "or," and "not." Strings of binary symbols, whether represented by zeros and ones or black and white marbles, could both encode and logically manipulate concepts of arbitrary complexity in unambiguous terms. This universal language would open a new era in human affairs. Leibniz saw Peter's ambitions as the means to propagate this revolution, drawing the analogy that building a new structure is easier than remodeling an old one whose foundations have settled unevenly, leaving defects that have to be repaired.

The Russian Academy of Sciences was founded in 1724. The Great Northern Expedition was launched in 1725, followed by a 126-year Russian presence in America, beginning with the arrival of Bering and Chirikov in 1741 and ending in 1867 with the transfer of Alaska to the United States. Leibniz's third project received no support. Although "so amused that he had looked at the instrument for half an hour," and even probed it with a pencil to see how it worked, Peter took no further interest in Leibniz's mechanical computer.[4] The powers of digital computing were lost on the tsar.

It took another two centuries, and the invention of electronics, for Russia, China, and the rest of the world to become the tabula rasa of Leibniz's plan. Then suddenly, in less than fifty years, we advanced from the first primitive electronic digital computers, assembled from vacuum tubes and exchanging coded sequences at the speed of punched cards and paper tape, to a world where code proliferates at the speed of light. The ability of digital computers to mirror the non-digital universe is taken for granted today. To question the supremacy of these powers elicits the same disbelief as trying to explain them did in the time of Peter the Great.

The differences between analog computing and digital computing are fundamental but not absolute. Analog computation deals with continuous functions, whose values change smoothly over time. Digital computation deals with discrete functions, whose values change in precise increments from one instant to the next. Leibniz might have envisioned an analog computer operating by means of a fluid running through a maze of pipes, regulated by valves that could be varied continuously between fully open and fully closed. As one of the founders of the infinitesimal calculus, he was no stranger to the continuous functions that such a device could evaluate or control. Instead, he envisioned a digital computer, with binary arithmetic executed by marbles shifted by on/off gates as they ran along multiple tracks.

These marbles were either black or white; no shades of gray allowed. They could not be divided into smaller marbles or merged into marbles of larger size. When arriving at a gate, they had to follow either one path or the other, with no middle ground. Any given sequence of marbles either corresponded exactly to some other sequence or did not. All questions had to be stated unambiguously, and if a question was repeated, the answer would be the same every time. This imagined computer was never built, but the binary digits, or bits, that permeate every facet of our existence are Leibniz's marbles, given electronic form.

Nature uses digital coding, embodied in strings of DNA, for the storage, replication, modification, and error correction of instructions conveyed from one generation to the next, but relies on analog coding and analog computing, embodied in brains and nervous systems, for real-time intelligence and control. Coded sequences of nucleotides store the instructions to grow a brain, but the brain itself does not operate, like a digital computer, by storing and processing digital code. "If the only demerit of the digital expansion system were its greater logical complexity, nature would not, for this reason alone, have rejected it," argued John von Neumann in 1948, explaining why brains do not use digital code.[5]

In a digital computer, one thing happens at a time. In an analog computer, everything happens at once. Brains process three-dimensional maps continuously, instead of processing one-dimensional algorithms step by step. Information is pulse-frequency coded, embodied in the topology of what connects where, not digitally coded by precise sequences of logical events. "The nervous system of even a very simple animal contains computing paradigms that are orders of magnitude more effective than are those found in systems built by humans," argued Carver Mead, a pioneer of the digital microprocessor, urging a reinvention of analog processing in 1989.[6] Technology will follow nature's lead in the evolution of true artificial intelligence and control.

Electronics underwent two critical transitions over the past one hundred years: from analog to digital and from high-voltage, high-temperature vacuum tubes to silicon's low-voltage, low-temperature solid state. That these transitions occurred together does not imply a necessary link. Just as digital computation was first implemented using vacuum tube components, analog computation can be implemented, from the bottom up, by solid state devices produced the same way we make digital microprocessors today, or from the top down through the assembly of digital processors into analog networks that treat the flow of bits not logically but statistically: the way a vacuum tube treats the flow of electrons, or a neuron treats the flow of pulses in a brain.

Leibniz's digital universe, despite its powers, remains incomplete, just as Isaac Newton, his rival over credit for the invention of the calculus, gave us a mathematical description of nature that predicts everything correctly, but only up to a certain point. The next revolution will be the coalescence of programmable machines into systems beyond programmable control.

There are four epochs, so far, in the entangled destinies of nature, human beings, and machines. In the first, preindustrial epoch, technology was limited to the tools and structures that humans could create with their own hands. Nature remained in control.

In the second, industrial epoch, machines were introduced, starting with simple machine tools, that could reproduce other machines. Nature began falling under mechanical control.

In the third epoch, digital codes, starting with punched cards and paper tape, began making copies of themselves. Powers of self-replication and self-reproduction that had so far been the preserve of biology were taken up by machines. Nature seemed to be relinquishing control. Late in this third epoch, the proliferation of networked devices, populated by metazoan codes, took a different turn.

In the fourth epoch, so gradually that almost no one noticed, machines began taking the side of nature, and nature began taking the side of machines. Humans were still in the loop but no longer in control. Faced with a growing sense of this loss of agency, people began to blame "the algorithm," or those who controlled "the algorithm," failing to realize there no longer was any identifiable algorithm at the helm. The day of the algorithm was over. The future belonged to something else.

A belief that artificial intelligence can be programmed to do our bidding may turn out to be as unfounded as a belief that certain people could speak to God, or that certain other people were born as slaves. The fourth epoch is returning us to the spirit-laden landscape of the first: a world where humans coexist with technologies they no longer control or fully understand. This is where the human mind took form. We grew up, as a species, surrounded by mind and intelligence everywhere we looked. Since the dawn of technology, we were on speaking terms with our tools. Intelligence in the cloud is nothing new. To adjust to life in the fourth epoch, it helps to look back to the first.

The beginning of this book is set at the close of the first epoch, the ending is set at the opening of the fourth, and the second and third epochs fall in between. The following chapters illuminate these transitions from a range of viewpoints over the past three hundred years. What drove the convergence of Leibniz's dreams of a digital universe with his mission to explore the American Northwest Coast? How did the two movements originate, and what led them to intersect? What are the differences between analog computing and digital computing, and why does this matter to a world that appears to have left analog computation behind? To someone who grew up in the third epoch but was drawn to the ways of the first, how to reconcile the distinction, enforced by the American educational system, between those who make a living with their minds and those who make a living with their hands? In an age that celebrates the digital revolution, what about those who fought for the other side?

The Bering-Chirikov expedition reached North America in 1741. The Russians, met by an indigenous population without written language but with advanced technology and arts, left a record of the Northwest Coast and its inhabitants at the moment that precontact times came to an end. Fifteen members of the expedition went ashore and were left behind. Their fate remains unknown.

At the close of the nineteenth century, the Chiricahua Apaches, descended from onetime Alaskans arriving from Asia who continued south, resisted subjugation to a later date than anyone else. In pursuit of the last of the Apaches, the U.S. government implemented the first large-scale high-speed all-optical digital telecommunications network in North America. The first shots in the digital revolution and the last bows and arrows deployed in war against regular soldiers of the U.S. Army were fired at the same time.

The invention of the vacuum tube, or thermionic valve, enabled machines with no moving parts except electrons, their operation limited not by the speed of sound that governs the transmission of information in mechanical devices but by the speed of light. It was into the war-surplus ferment of the electronics industry that otherwise abstract contributions from theoretical physics and mathematical logic combined to realize Leibniz's vision of binary arithmetic as a universal code. The vacuum tube, treating streams of electrons as continuous functions, was an analog device. The logical processing of discrete pulses of electrons had to be imposed upon it, against its will, until the advent of the transistor brought this age of reptiles to a close.

The Hungarian physicist Leo Szilard, after helping to invent nuclear weapons, spent the rest of his life opposing them—except for their use in the exploration of space. This possibility was taken up by a privately organized, government-supported project whose mission was to reach Saturn by 1970 in a four-thousand-ton spaceship carrying one hundred people on a voyage modeled after that of Darwin's *Beagle*, allowing four years to complete the trip. Project Orion was abandoned by the U.S. government, while Szilard's fictional *Voice of the Dolphins* led to my own adventures on the Northwest Coast.

Three of those years were spent in a tree house ninety-five feet up in a Douglas fir above Burrard Inlet in British Columbia, on land that had never been ceded by its rightful owners, the Tsleil-Waututh. Trees integrate a range of continuous inputs into a single channel of digital output: growth rings that are incremented one year at a time. I was surrounded by growth rings going back to the year 1426.

My own version of string theory holds that lashing and sewing are overlooked drivers of our technological advance. On the Northwest Coast,

Russian American colonists adopted the indigenous technology of the skin-boat builders rather than replacing it with something else. I took the Russian adoption of the Aleut kayak, or *baidarka*, as a model, not only for my own boatbuilding but also for how technology is emulating the design and tactics of living organisms on all fronts.

Samuel Butler's "Darwin Among the Machines," appearing out of nowhere in the New Zealand wilderness of 1863, was fleshed out into his prophetic, dystopian *Erewhon* of 1872. In his notes for *Erewhon Revisited*, Butler went on to warn us that the advance of artificial intelligence would be driven by advertising and that both God *and* Darwin might turn out to be on the side of the machines.

The optically transmitted intelligence and numbered identity tags of the nineteenth-century campaign against the Apaches have descended to the optically fed data center recently established in the nearby desert by the National Security Agency. In the analog universe, time is a continuum. In the digital universe, time is an illusion conveyed by sequences of discrete, timeless steps. No time is there. What happens to Leibniz's vision of a digital enlightenment when all human activity is machine-readable and all steps can be retraced?

In 1890, after the exile of the Chiricahua Apaches as prisoners of war to Florida, a vision received by the Paiute prophet Wovoka led to a grassroots movement among the North American First Nations, promising to bring their dead warriors and dying ways back to life. An analogous prophecy, conveyed through a mathematical conjecture known as the continuum hypothesis, suggests that the powers of analog computation, transcending the bounds of algorithms, or step-by-step procedures, will supervene upon the digital and reassert control. Electrons, treated digitally, became bits; bits, treated statistically, have become electrons. The ghost of the vacuum tube has returned.

Leibniz's ideas arrived in North America twice: in the twentieth century with the digital computer, and with the Bering-Chirikov expedition in the

eighteenth. When the navigators following Peter's instructions reached
the American Northwest Coast, they were met by people who had been
doing just fine since the last technological revolution, some fifteen thousand years earlier.

The slate was not blank.

1741

Nikita Shumagin, seaman second rank, was the first to die of scurvy, on August 30, 1741.[1] He had been carried ashore in the hope that fresh air and water would save his life but died within hours, the first of a series of fellow crewmen who "died like mice as soon as their heads had topped the hatch."[2] He was buried in a shallow grave on Nagai Island, one of a cluster of islands off the Alaska Peninsula known as the Shumagin group today. His death left seventy-six men and an eleven-year-old boy to make their way back to Kamchatka from America in a disintegrating eighty-foot boat.

Georg Wilhelm Steller, the expedition's naturalist, knew that the treeless island's abundance of low-lying vegetation could restore the crew to health, but his advice was ignored. "I asked for a few men to gather up as many antiscorbutic plants as we would need," he complained, but "the gentlemen scorned even this." Steller alone gathered scurvy grass (*Cochlearia*), *Lapathum*, and other greens onshore while the ship's medicine chest remained "filled with the most useless medicines, almost nothing but plasters, ointments, oils, and other surgical supplies needed for four

View *of the* Islands *and* Main Land *on the* N *of* Schumagin's Strait *when passing them*.

to five hundred men with wounds from great battles, but with nothing whatever needed on a sea voyage where scurvy and asthma are the chief complaints."[3]

Steller, the sole representative of Science on board the ship, was a late addition to a group of academicians who had left St. Petersburg in August 1733, attached to a voyage of discovery set in motion by Peter the Great in 1724, just months before his death. The vast undertaking, whose fate now lay in the hands of a thirty-two-year-old botanist walking alone along a windswept shore, had originated with Leibniz's proposal to the tsar. Because Russia had no Pacific fleet with which to undertake the voyage to America, it would be necessary to build one. Nothing excited the tsar as much as building ships. If he were younger, he would have led the voyage himself.

Peter had been proclaimed tsar in 1682 at the age of nine. Three weeks later, he had watched from the palace balcony as his maternal uncle Ivan Naryshkin and his protector Artamon Matveev were hacked to death by a pike-wielding mob. He was left with a tendency to brutality and distrust that culminated, in 1718, with the torture and sentencing to death, on charges of treason, of Alexis, his own son. At fifteen he was given an astrolabe by Franz Timmerman, a Dutch seaman attached to the royal household, and instructed in its use. At sixteen, at the royal estate at Izmailovo on the outskirts of Moscow, he discovered a small sailing vessel left in storage and disrepair.

"It happen'd that his Majesty was in the Flax-yard at *Ismaeloff*, and walking by the Magazines, where some Remains of the Household Furniture of *Niketa Ivanowich Romanoff*, his Great Uncle were laid," the preface to the Russian Naval Statute of 1720 explains. "He espy'd amongst other Things a small foreign vessel, and his native Curiosity not suffering him to pass it by without an Enquiry, he presently asked *Francis Timerman* (who then liv'd with him and taught him Geometry and Fortification) what was that? He told him it was an *English* Boat . . . [and] that it goes with a Sail, with a Wind, or against it; which Word made him greatly Wonder; and as tho' not credible, rais'd his Curiosity to see a Proof of it."[4] Timmerman and a fellow shipwright, Karsten Brant, restored the vessel and demonstrated that it could, unlike existing Russian vessels, be sailed upwind. Peter was hooked. He became obsessed with shipbuilding and seafaring despite there being only one seaport, at Arkhangelsk on the White Sea, in Russia at that time.

Peter supervised the construction of modern shipyards, founded the Russian navy, and established St. Petersburg, on the Baltic, as the headquarters of the Russian fleet. He led training exercises, launched naval battles against the Turks and Swedes, and instituted a penalty of five rubles per oar for rowing on the river Neva if there was enough wind to sail. In 1697 he traveled incognito to Holland and worked as a shipwright in the East India Company yards, followed by a three-month stay in London, where he and his entourage rented the manor house on Sayes Court owned by John Evelyn, adjacent to the king's shipyard at Deptford on the Thames. Evelyn, who at first welcomed the tsar's "having a mind to see the building of ships," began to have second thoughts upon receiving a report that his house, featuring a private entrance to the dockyards, was "full of people, and right nasty. The Czar lies next to your library, and dines in the parlor next to your study." Peter was "seldom at home a whole day" and spent most of his time in the shipyard, or out on the water, wearing commoner's dress.[5]

"He's a great admirer of such blunt fellows as saylors are," Thomas Hale, an English merchant, confirmed. "He invited all the nasty tars to dinner with him where he made 'em so drunk that some slop't, some danced, and others fought—he amongst 'em . . . None can complain of his frolicks since he himself is allways the first man."[6] By the time Peter departed on April 21, 1698, the damages to Evelyn's premises were so severe that Sir Christopher Wren, who had helped direct the reconstruction of London after the fire of 1666, had to be brought in to estimate the cost of repairs.

Mineral water replaced brandy and vodka during Leibniz's audience with the tsar. "I spent almost eight full days at the mineral baths in Pyrmont

attending the great Russian monarch," Leibniz reported to Johann Bernoulli, "and the more I observe the character of this prince, the more
I admire him." The tsar's inclinations to experimental science, Leibniz
explained, were evidenced by his drawing blood samples from all the
members of his party before and after taking the cure, "including the sole
Russian priest among them, whose pale, thick blood was the worst of all."
Peter, fascinated by medical procedures, was known to perform minor
surgery, including tooth extractions, himself. "After the time of drinking
had ended," Leibniz continues, "the prince, ingenious as he is, decided to
conduct an experiment to determine the effects of the waters, and had the
priest's vein pierced again. The blood drawn was the purest red and of the
sort you would expect in the healthiest person, I myself was there when it
was brought out. The prince applauded, and not without reason, for such
a remarkable change in such a short amount of time could hardly be attributed to diet alone."[7]

Leibniz was awarded a salary of a thousand thalers a year and a secret
diplomatic mission to Vienna by the tsar. He contributed to projects ranging from the founding of the Russian Academy of Sciences to building a
mechanical support for the tsar's paralyzed arm, and appeared destined to
play a growing role in Peter's court. "I am to become in some way the So-
lon of Russia," he wrote to Sophie, the electress of Hannover, explaining
that "the shortest laws like the ten commandments of God and the twelve
tables of ancient Rome are the best."[8] In Leibniz's philosophy, our universe had been selected from an infinity of possible universes so that the
minimum number of natural laws would produce the maximum diversity
of results. The instructions that launched the American expedition were
Leibnizian in brevity and scope:

1. In Kamchatka or some other place build one or two boats with
   decks.
2. On those boats sail near the land which goes to the north (since
   no one knows where it ends) it seems is part of America.
3. Discover where it is joined to America, and go as far as some

town belonging to a European power; if you encounter some European ship, ascertain from it what is the name of the nearest coast, and write it down and go ashore personally and obtain firsthand information, locate it on a map and return here.[9]

These orders were issued by Catherine I upon Peter's death in January 1725. No expense was to be spared. Thus Steller, sixteen years later, found himself gathering beach greens in hopes of keeping the crew alive to bring back the promised reports. The traveling academy that had left St. Petersburg eight years earlier included the astronomer Louis Delisle de La Croyère, the botanist Johann Georg Gmelin, and the historian and ethnographer Gerhard Friedrich Müller, as well as two artists, a surgeon, an interpreter, an instrument maker, five surveyors, six scientific assistants, and fourteen bodyguards, and was "right luxuriously equipped."[10] They carried an extensive library and were granted the authority to appropriate local accommodations as necessary to assist the professors in their pursuits. La Croyère alone left St. Petersburg with nine wagonloads of instruments, including telescopes thirteen and fifteen feet in length. Gmelin's party was equipped with several barrels of a particular German wine. The academic entourage occupied twelve riverboats, outfitted with cabins for the professors, on the voyage up the Lena River to Yakutsk, the limit of established navigation, where these comforts had to be left behind.

The Great Northern Expedition, or Second Kamchatka Expedition as it became better known, was the most extravagant geographic exploration ever launched. It was led by the Dane Vitus Bering, assisted by the Russian Alexei Chirikov, and lasted nine years, following the First Kamchatka Expedition, also led by Bering and Chirikov, which had lasted six. More than one thousand individuals were assigned to the second expedition, augmented by at least two thousand exiles and other subjects recruited along the way.

The burden of transporting supplies and equipment six thousand miles across Siberia to the frontier seaport of Okhotsk was shared between animals and men. Pack horses carried 180-pound loads through the mud in summer, and human conscripts dragged sledges in winter along frozen trails. Both horses and men were worked to death. "As our labourers, that is to say the deported persons, were deserting in great numbers," explains Sven Waxell, Bering's lieutenant and father of the eleven-year-old Laurentz, "we sought to prevent further losses by introducing harsh discipline; we set up a gallows every twenty versts [thirteen miles] along the river Lena, which had an exceptionally good effect."[11]

The expedition requisitioned 4,280 pairs of saddlebags, consumed 180,000 pounds of rye flour per year, and descended upon the local population like a plague of locusts as it made its way to the Pacific coast. The main body of the expedition took three years to make the trip. To proceed beyond Yakutsk, hundreds of crude riverboats were built. They were hauled up the Aldan, Maya, and Iudoma Rivers by men trudging along the insect-infested, hazard-encumbered banks, spending months working their way upstream over distances that could be covered in days on the way down. A final overland leg extended through uninhabited wilderness from the headwaters of the Iudoma to the headwaters of the Urak. "We had four or five hundred men continually on the march to and fro," Waxell reported after transporting six hundred tons of equipment and supplies over this route in 1737, "the whole time harnessed to their sledges like horses [with] only what the country could provide, which was rye-flour and some groats."[12]

At the terminus of the Iudoma portage, a fleet of even cruder vessels was built, braving the rapids by which the Urak reached the Sea of Okhotsk, where the ships were broken up and burned for firewood if they survived the descent. The heavy freight included the ships' anchors as well as eighteen cannons weighing 738 pounds each, ten cannons weighing 666 pounds each, and 14,400 cannonballs, all cast in central Siberia by a foundry at Kamenskii near Tobol'sk—still thirty-five hundred miles from Okhotsk. After seven years of struggle, two seagoing vessels, modeled on the Dutch packet boats that had so captivated Peter during his apprenticeship, were completed at the improvised shipyard in Okhotsk, and on September 8, 1740, the voyage to Kamchatka and America began. Bering was in command of the *St. Peter* and Chirikov was in command of the *St. Paul*.

Gmelin and Müller abandoned the expedition before its departure from Kamchatka for America, while La Croyère, who sailed with Chirikov aboard the *St. Paul*, failed to secure any astronomical observations but managed to fend off the scurvy, leaving it "wondered," as it was anonymously reported, "that the great quantity of brandy which he swallowed every day had such a good effect." He survived the voyage, only to die upon arrival back in Kamchatka, "having dressed himself in order to go ashore, and having given once more an extravagant vent to his joy at his safe return."[13] Of the academicians, only Steller, an adjunct member of the academy recruited to the expedition in 1735 at the age of twenty-six, accompanied by a single Cossack rifleman serving as his assistant and unencumbered beyond his own personal effects, returned alive from America with observations, specimens, and notes.

Bering was a broken man by the time the expedition left Kamchatka on June 4, 1741, rarely leaving his cabin and deferring to Lieutenant Waxell and the fleet captain Sofron Khitrov, who supervised the operation of the ship. The *St. Peter* and the *St. Paul* became separated in a storm on June 20 and were never rejoined. Both vessels fired their cannons at intervals according to the predetermined code of signals, but after three days with no answer from an empty sea they gave up and continued eastward on separate paths into the unknown. The *St. Peter* made landfall off Cape St. Elias at Kayak Island on July 16, 1741, dropping anchor in the lee of the island early in the morning of the twentieth, when Steller, who had seen land first, was allowed to go ashore.

He found campfires that were still warm, stones showing signs of having been used to sharpen copper knives, and a subterranean cache of food, including smoked salmon "so cleanly and well prepared that I have never seen it as good in Kamchatka, and it was also much superior in taste."[14] A sample was delivered to Commander Bering, who ordered that a dozen yards of green Chinese silk, two iron knives, an iron kettle, and twenty strings of glass beads be taken ashore and left in its place. "In exchange," Steller reported, the cache "was plundered to such an extent that if in the future we were to return to this place, the people would flee from us just as they did this time."[15]

Steller had only begun to explore the island when Khitrov ordered him to return to the ship. The next morning, "two hours before daybreak, the Captain Commander, much against his usual practice, got up and came on deck and, without consulting anyone, gave orders to weigh anchor."[16] Bering, who would not live to see it, set sail for home. The expedition never set foot on the American mainland. So much for the instructions left by the tsar. "Ten years the preparations for this great undertaking lasted," Steller complained, "and ten hours were devoted to the work itself."[17]

Bering left his name to an island, a sea, and a strait; Steller left his name to a sea cow, a sea lion, and a jay. Steller cultivated an adversarial relationship with the ship's officers, especially Khitrov, whose contributions to their misfortunes began at the outset with the loss of the expedition's entire consignment of ship's biscuit, being ferried to Kamchatka from Okhotsk. The *Nadezhda*, a shallow-draft supply vessel under Khitrov's command, ran aground in leaving the harbor, an omen of things to come.

A replacement cargo of biscuit had to be sent overland across the Kamchatka Peninsula by sledges during the winter, using some five thousand dogs commandeered from the Kamchadals. Seven members of a contingent of Russians sent to secure the dogs, which their owners "loved above all things," according to Waxell, were asphyxiated by flaming kindling dropped into the semi-subterranean lodgings they had expropriated from the Kamchadals.[18] The Russians retaliated indiscriminately, with grenades. The delay over the loss of the biscuits led to the expedition's wintering in Kamchatka, rather than in America as had been planned.

By September, when the *St. Peter* stumbled upon the Shumagin Islands, the remaining biscuits were spoiled and freshwater had run out. Steller discovered an upland spring, but Khitrov ordered the empty water casks filled from a brackish lagoon where they landed their boats. According to

Russian naval custom, critical decisions required a sea council among the ship's officers, with any dissenting opinions noted and taken into account. But the council had long ceased listening to Steller, not for being wrong, but for always believing he was right. "These gentlemen want to go home, and that by the shortest road but in the longest way," he complained as they set off with contaminated water on what he argued was the wrong course.[19] Steller blamed Khitrov for reducing their voyage of discovery from the collection of knowledge to the collection of drinking water from America at the highest possible cost.

In making their way back out to sea, the *St. Peter* was forced to seek refuge in the lee of Bird Island, where, on September 4, 1741, "without expectation or search we chanced to meet with Americans," Steller wrote. "We had scarcely dropped the anchor when we heard a loud shout from the rock to the south of us, which at first not expecting any human beings on this miserable island twenty miles away from the mainland, we held to be the roar of a sea lion. A little while later, however, we saw two small boats paddling toward our vessel from shore."[20]

The two paddlers, each in a skeletal, skin-covered kayak termed a "baidarka" by the Russians, approached the ship and then "began, while still paddling, simultaneously to make an uninterrupted, long speech in a loud voice of which none of our interpreters could understand a word." One of the paddlers, after "paint[ing] himself from the wings of the nose across the cheeks" with shiny mineral pigment, took one of a bundle of spear shafts, about four and a half feet long, that were painted red and stowed on the aft deck of the kayak, "placed two falcon wings on it and tied them fast with whalebone, showed it to us, and then with a laugh threw it towards our vessel into the water." The Russians responded by placing some glass beads and two Chinese tobacco pipes on a board and tossing it into the water, the gifts being accepted and placed on the deck of the other kayak. The first paddler then "became somewhat more courageous, approached still nearer to us, though with the greatest caution, tied an eviscerated entire falcon to another stick and passed it up to our Koryak interpreter in order to receive from us a piece of Chinese silk and a mirror."[21]

The islanders beckoned the visitors to follow them ashore. Waxell, Steller, and the Kamchadal interpreter Aleksei Lazukov, accompanied by nine sailors and soldiers equipped with lances, sabers, and guns concealed under a canvas tarp, lowered the longboat and followed the kayakers but were unable to land because of rocks and breaking waves. Lazukov and two of the Russians took off their clothes and waded through the surf to shore. They were received "quite deferentially as if they were very great personages," treated to whale blubber, showed where they could obtain water, and engaged for an hour in attempted conversation with their hosts. Steller and Waxell remained in the longboat, where they were visited by the apparent leader of the islanders who had returned to his kayak, "which he had lifted with one hand and carried under his arm to the water, and came paddling up to us." The emissary "was made welcome with a cup of brandy, which, following our example, he emptied quickly, but also immediately spit out again."[22]

After two hours, with the wind rising, the landing party was recalled, but the islanders tried to persuade the visitors to stay, first with gifts and then by holding Lazukov and attempting to pull the longboat into shore. "Nine of the Americans seized him and would not let him go," Khitrov reported, "which shows that they regarded him as one of their own people even though he is a Kamchadal."[23] The crew in the longboat then fired two muskets in the air, at which the islanders released the interpreter and fell to the ground, allowing the Russians to make their escape. The islanders then "waved their hands to us to be off quickly as they did not want us any longer," Steller reported. "Some of them in getting up picked up stones and held them in their hands."[24]

This was the first recorded contact by Europeans with a people whose presence among these islands went back at least ten thousand years. Fifteen thousand years ago, with the Pleistocene ice sheet in retreat, sea level was a hundred meters lower than today. A diminished Bering Sea bordered the southern shore of Beringia, a broad, low-elevation landmass joining Asia to America that was six hundred miles wide.

Humans arrived in America while much of the ice sheet was still in place, either through an inland ice-free corridor or along the shoreline, or both. The coastal route now lies submerged, but growing evidence supports the "kelp highway" hypothesis of Knut Fladmark and Jon Erlandson: that the first arrivals spread eastward along the edge of Beringia and then rapidly moved south.[5] Even during the glacial maximum there were ice-free refugia along the outer coast. Shellfish were available at low tide, shorebirds gathered by the millions, and sea mammals that hauled out on land would have been easy prey.

These first Americans either left Asia as skin-boat builders or became skin-boat builders along the way. Once sea mammals were disassembled for food, fuel, and shelter, only a driftwood skeleton was needed to reassemble them as skin boats. At least one branch of migrants, remaining in the Aleutian Islands, became the Unangan, whom the Russians designated, along with some of their immediate neighbors, as "Aleuts."

The Aleutian Islands are treeless not from cold but from wind. The landscape is blanketed with chest-high grass, while the wind that prevents the growth of forests has piled the beaches high with wood. The Aleut answer to these conditions was the construction of semi-subterranean houses and semi-submersible boats. Large communal dwellings and almost unlimited marine resources supported population densities atypical of preagricultural times. Anangula, a settlement on what was once the end of a peninsula and is now a small island just off Umnak Island at the eastern end of the Aleutian chain, shows a record, based on the excavation of more than two million artifacts, of continuous occupation by sea mammal hunters going back ninety-eight hundred years. In the words of the anthropologist William S. Laughlin, who began investigating the Anangula site in 1938, it was the physical and intellectual demands of open-water sea mammal hunting that led the Aleuts "to shuffle a whole lot faster through the evolutionary deck."[26]

The Aleutian Islands, extending for fifteen hundred miles, are close enough together that it is possible to paddle from one island to the next, but far enough apart that exceptional skills were required to populate the chain. Local cultures and kayak designs could differentiate, but no one group became isolated from the rest. There was continuous competition, intermittent warfare, and cross-pollination of design. Hundred-mile open ocean crossings and whale hunting by solo kayakers armed with poison-tipped harpoons became routine. The Aleuts, as amphibious as the animals on which they preyed, were as integrated with their kayaks as the Apaches and Comanches would become integrated with the wild ponies they adopted as mounts.

Unangan kayaks, termed *iqyax̂* (with one hatch) and *ulux̂tax̂* (with two hatches), evidenced innovations found nowhere else: articulated skeletons, a compound wide-tailed stern, bifurcated surface-piercing bows, and leaf-shaped paddles having complex lift-producing surfaces similar to the "wing" paddles developed for racing kayaks in recent years. "With this he beats alternately to the right and to the left into the water and thereby propels his boat with great adroitness even among large waves," noted Steller after the first encounter with Aleuts. The paddler wore a waterproof jacket sewn in a continuous helix from a strip of oiled whale intestines and sealed by a secondary gasket to the hatchway around his waist. "Even in very rough weather . . . when really high seas are running," observed Waxell, "once seated in the kayak and thus fastened in it, not a drop of water can find its way inside."[27]

The *St. Peter* was treated less kindly by the waves. Bering's crew clawed their way westward against unrelenting headwinds, described in the otherwise matter-of-fact logbook as "terrific" and "frightful," with seas washing over the decks. Sails and rigging began to fail. The log of the *St. Peter*, kept by the assistant navigator Kharlam Yushin, notes that at five o'clock in the afternoon on September 23 "by the will of God died of scurvy the grenadier Andrei Tretyakov," and at eleven o'clock in the morning, on the twenty-fourth, they "lowered the dead body into the sea."[28]

"People's minds again became as loose and unsteady as their teeth were from scurvy," Steller noted on October 2, and on October 10 the rain began to turn to snow.[29] "The disease first showed itself in a feeling of heaviness and weariness in all our limbs, such that we were all the time wanting to sleep and, having once sat down, were most reluctant to rise again," Waxell added. "We became more and more depressed." The supply of vodka ran out. "As long as it had lasted, it had kept the men in fairly good fettle," he notes. "Deaths now became numerous, so much so that a day seldom passed without our having to throw the corpse of one of our men overboard."[30]

By October 13, twenty-one men were incapacitated, and on October 19 "by the will of God Alexei Kiselev died of scurvy; 29 men on the sick list." The next day "Nikita Kharitonov died by the will of God," and two days later "by the will of God died the marine Luka Zaviakov. We wrapped the dead marine and dropped him into the sea." The log entries then shift to the first person, noting that "today I became very ill with the scurvy" and on the twenty-seventh "I can with difficulty stand my watch."[31]

Between November 3 and November 4 there were three deaths in seventeen hours. Those still alive were giving up all hope when, at eight o'clock in the morning on the fourth, land was sighted to the north. "It is impossible to describe how great and extraordinary was the joy of everybody at this sight," notes Steller. "The half-dead crawled up to see it [and] little cups of brandy concealed here and there made their appearance in order to keep up the joy."[32]

In deciding whether they should keep trying to reach Avacha, the harbor they had sailed from five months earlier, or head for the land in front of them, which Khitrov swore could not be anything except Kamchatka, "we discovered that all our big mainstays on the starboard side had parted, nor did we have anyone able to repair the damage," says Waxell. Inspecting further, they found that "the main shrouds had likewise parted on the larboard side."[33] Forty-nine of the crew were now incapacitated, leaving only fourteen with any ability to manage the ship. Bering, who advocated pressing on for Avacha, convened a council in his cabin for a brief and one-sided discussion, dominated by Waxell and Khitrov, with the lone voice of dissent, besides Steller's, dismissed from the cabin by Khitrov's exclaiming, "Out! Shut up, you dog, you son of a bitch!"[34]

Determined "to make for the land we had seen and by that means seek to save our lives," they approached the shore under shortened sail and attempted to anchor overnight, but another storm came up.[35] Both main and reserve anchor cables parted, with huge waves driving the ship toward the rocks. They could do nothing more to save themselves except to enlist the help of their dead companions who were now "flung without ceremony head over heels into the sea," notes Steller, "since some superstitious persons, at the start of the terror, considered the dead as the cause of the rising sea."[36] With the corpses cast to the waves, the ship was swept over the reef intact, leaving them secured, with the last of their reserve anchors, in four and a half fathoms of water "as in a placid lake."[37]

November 7 dawned calm and bright. Steller, his Cossack servant Thoma Lepekhin, the two Waxells, Bering's clerk and expedition artist Friedrich Plenisner, and several others prepared to go ashore. They expected to find the nearest settlement, arrange for horses to transport the invalids to safety, and enlist a fresh crew to preserve the ship. "We were not yet on the beach when something struck us as strange," Steller reported, "namely, some sea otters came from shore toward us into the sea."[38] In Kamchatka, sea otters avoid people and are never found on land.

Plenisner shot half a dozen ptarmigans, sending a portion of them, along with a salad collected by Steller, back to Commander Bering with Lieutenant Waxell, while Steller made the remainder into soup for the shore party who spent the night huddled in an improvised driftwood hut. Bering, Khitrov, and most of the survivors soon moved ashore, occupying crude dugouts carved out of the sand and covered with salvaged pieces of sail. Waxell remained on board with twelve living crew members and five dead, battered by seas that "poured into our ship over the sides." They abandoned ship on November 20, finding life little better ashore. "Our plight was so wretched that the dead had to lie for a considerable time among the living, for there was none able to drag the corpses away," wrote Waxell, "nor were those who still lived capable of moving away from the dead."[39] Hordes of blue foxes "ate the hands and feet of the corpses before we had time to bury them."[40] Steller and Plenisner killed sixty foxes in a single day, using the carcasses to seal the gaps in the walls of their hut.

On November 23 the Sea Council resolved to save the *St. Peter*, or at least salvage its remaining contents, by bringing the ship ashore, but they could not muster enough men to raise the anchor and gave up. Five days later, after an overnight storm, they found the ship had beached itself, "better than human industry could perhaps have accomplished it," in Steller's account.[41]

There were twenty-one deaths in November and December, including Bering, who "died miserably under the open sky on December 8th, almost eaten up by lice."[42] Fresh food and water came too late to save his life. "The Captain-Commander lay in a little hollow in the sand," Waxell explains. "The sand kept trickling down the sides and in the end had filled the hollow about half full. As the Captain-Commander was lying in the centre of the hollow, it ended by the half of his body being covered by the falling sand . . . Even if it had been possible to pull him out again, it would have been against his wishes, for he said to us: 'The deeper in the ground I lie, the warmer I am; the part of my body that lies above ground suffers from the cold.'"[43]

Bering was reburied in a crude coffin where his remains rested undisturbed until exhumed in 1981 by a team of Russian archaeologists, who also recovered seven of the *St. Peter*'s fourteen cannons, concealed under sand and gravel where the ship had come ashore. The heavy armament that had cost so many lives to transport across Siberia to Kamchatka and America had been used only for signaling and had never fired a hostile shot.

Steller now flourished. Marooned on an uninhabited island, he came into his own. The fourth of ten children, he had been trained first as a Lutheran minister, then as a physician, and finally as a botanist. Officially, he was still the expedition's mineralogist, assigned to report on any geological resources discovered in America or along the way. Because the original botanist, the designated ethnographer, the chief physician, and even the official clergyman had all abandoned the voyage, he assumed those responsibilities as well. His dietary prescriptions restored the survivors to health, while his faith in God was mixed with practical enterprise including the collection of a fortune in sea otter pelts. He came to be as trusted by the shipwrecked as he had been distrusted on board the ship. Old feuds were put aside, although he refused a place in his hut to Khitrov and cared for Waxell partly because of his fondness for the eleven-year-old Laurentz and partly because if Waxell, the acting commander, had died, Khitrov would have assumed command. Steller, says Waxell, "showed us many green herbs . . . and by taking them we found our health noticeably improved. None of us became well or recovered his strength completely before we began eating something green."[44] With the death of the petty officer Ivan Lagunov on January 8, the toll from scurvy came to an end.

"We all realized that rank, learning, and other distinctions would be of no advantage here," Steller noted, and "began to regard many things as treasures to which formerly we had paid little or no attention, such as axes, knives, awls, needles, thread, shoe twine, shoes, shirts, socks, sticks, strings, and similar things which in former days many of us would not have stopped to pick up."[45] To build morale among the crew, who could have mutinied at any time, it was decided to "address everybody somewhat more politely by their patronymics and given name [and] we soon learned that Peter Maximovich was more ready to serve than Petrusha was formerly."[46] They divided their time between retrieving driftwood for fuel, hunting animals for food, foraging for greens, and exploring their surroundings, which they soon confirmed were an island, named Bering Island after Bering's death.

Bering Island was an island in time: a living relic of what the first to venture along the shores of Beringia might have found. Sea otters "covered the shore in great droves," Steller noted, "so far from fearing man that they would come up to our fires and would not be driven away until, after many of them had been slain, they learned to know us and run away." The castaways collected more than nine hundred pelts and thanks to the meat "were saved from scurvy, and no one got sick of it, although we ate it every day half raw."[47] Steller had learned of antiscorbutic plants from the Kamchadals, who also might have known that cooking freshly killed meat would destroy the vitamin C that it contained. With no predators on the island, sea otters roamed onshore, even ascending to the freshwater lakes in the interior of the island, from where "on warm days they seek the valleys and shady recesses of the mountains and frolic there like monkeys."[48] In early spring, fur seals began showing up, and on April 20 a whale, ninety feet long, washed ashore. "By means of sea animals," mused Steller, "it pleased God to strengthen us human beings who had been shipwrecked through the sea."[49]

On April 9 it was decided to break up the *St. Peter* and construct a smaller vessel from the remains. The keel was hewn from the mainmast of the *St. Peter* and laid on May 6. To feed the workers and stockpile provisions for the continuation of the voyage, the hunters began to harpoon and butcher some of the sea cows grazing in the kelp beds along the shore. "When the tide came in they came up so close to the shore," notes Steller, "that I sometimes even stroked their backs with my hand."[50] This population, like the last herd of woolly mammoths that survived on Wrangell Island until thirty-seven hundred years ago, was a remnant of the age of megafauna that the arrival of humans had brought to a close. "There is so large a number of these animals about this one island that they would suffice to support all the inhabitants of Kamchatka," Steller wrote.[51]

Adults were thirty to thirty-five feet long and weighed about eight thousand pounds. They fed primarily on kelp, chewed between masticatory plates before entering a digestive tract that measured 497 feet long "from gullet to anus" in the case of an adult female that Steller dissected, whose stomach was "of stupendous size, 6 feet long, 5 feet wide, and so stuffed with food and seaweed that four strong men with a rope attached to it could with great effort scarcely move it from its place and drag it out."[52] Steller had to pay his anatomical assistants out of his own personal supply of tobacco to secure their help.

The sea cow (*Hydrodamalis gigas*) was doomed to extinction. Its meat was of "excellent taste, not easy to distinguish from that of beef," while that of the calves was "just like veal" and its fat "so sweet and fine flavored that we lost all desire for butter." Its hide would become favored for covering large skin boats, and its oil, used in lamps, "burns clear, without smoke or smell." The meat was readily preserved, and "even in the hottest days it can be kept in the open air for a very long time."[53] The Bering Islanders found that a single sea cow would feed the entire party for two weeks, and "all who ate it felt that they increased notably in vigor and health."[54]

On August 9, 1742, the derivative *St. Peter*, thirty-six feet long, was skidded over the abandoned cannons, laid out to form a slipway, into the water at high tide. The forty-six survivors, so cramped that they had to sleep in turns, left Bering Island on August 13 and reached Avacha Bay on August 26, where they were greeted by a Kamchadal paddler who informed them they had been declared dead, their salaries had been terminated, and all the property they had left behind had been sold.

Laurentz Waxell was now twelve.

Steller unloaded his collections at Petropavlovsk, including his personal cache of some three hundred sea otter pelts, before making his way on foot across the Kamchatka Peninsula to Bol'sheretsk. "Nobody was luckier than Steller," reported Müller, who had abandoned the expedition but was the first to publish an account. "As a doctor he received many skins as presents, and many others he acquired from those who, in the uncertainty over whether they would ever again meet with men, did not value these goods."[55] The furs would fetch thirty to forty rubles each in Yakutsk.

Steller soon fell into a bitter dispute with Vasili Khmetevski, the naval administrator of Bol'sheretsk, whom he accused of mistreating the Kamchadals. Khmetevski accused Steller, who had used his influence to secure the release of some innocent Kamchadals held as prisoners, of fomenting rebellion, the charges and countercharges being forwarded to the Senate in St. Petersburg by the same post. After two more years exploring Kamchatka and the Kuril Islands, including an unsuccessful attempt to secure a killer whale for his collection of marine mammals, Steller began the long journey back to St. Petersburg, with sixteen crates of specimens, in August 1744. He arrived in Yakutsk in October, spent the winter there, left when the ice on the Lena broke in May 1745, and reached Irkutsk by fall. He continued to advocate on behalf of his native informants, or, as Müller put it, he now "immerged himself without necessity, though with a good intention, in matters that did not belong to his department," by filing a new series of complaints.[56]

Steller was placed under arrest, to the satisfaction of his nemesis Khmetevski, in December 1745, but was soon exonerated by the vice-governor in Irkutsk, whom he then denounced, during a night of heavy drinking, with the words "slovo i dielo" (in word and deed), an accusation of treason requiring both parties to appear before the tsar's secret tribunal in Moscow, where statements would be extracted by torture under the knout. Steller was allowed to retract his accusation in the morning and left by sledge and post-horse on Christmas Eve, but the news of his acquittal on Khmetevski's charges did not reach St. Petersburg in time to prevent orders being issued to intercept him as he left Siberia and return him to Irkutsk.

On August 16, 1746, the Senate's order caught up with him in Solikamsk, where he was given only twenty-four hours to secure his latest botanical collections for shipment and storage before being escorted, "with only one coat and 60 rubles,"[57] back into the depths of Siberia. He and his guard, who was as unhappy with the assignment as he was, had reached Tara and the onset of Siberian winter in early October, when a Senate courier arrived by troika with orders to rescind his arrest. Steller reversed course, in bitter cold, toward the west. He stopped to celebrate, drinking too heavily, in Tobolsk, where he took ill with a high fever but insisted on pushing on until forced to stop and seek medical attention in Tiumen, Siberia, where he died on November 12, 1746.

Steller's shallow grave, in frozen ground, was soon plundered by robbers and wild dogs. His only lasting monument was the body of work that arrived at the academy in St. Petersburg after his death. "I do not doubt that the American shores are to become better known to us," he predicted in his monograph on marine mammals, written during his stay on Bering Island and published posthumously in 1751.[58] It was his own collection of sea otter pelts that precipitated the rush to Alaska by fur hunters who exterminated the sea cow, almost exterminated the sea otter, and brought the same subjugation to the Aleuts from which he had tried to protect the Kamchadals.

The Steller sea cow was last seen alive in 1768, after a series of expeditions revisited Bering Island to hunt the animals in bulk. "As long as things escape us and perish unknown with our consent, and through our silence," Steller, the sole naturalist to ever observe the animal, warned, "it is not strange that these things, which we are prevented from observing by the great sea that lies between, have remained to the present time unknown and unexplored."[59]

The voyage to America unfolded much differently aboard the *St. Paul*. On July 15, 1741, just one day before Steller and the *St. Peter* made landfall off Cape St. Elias, Chirikov and the *St. Paul* made landfall near Cape Addington, five hundred miles to the southeast. Following the heavily forested, mountainous coastline northwestward, they prepared to send a party ashore. Like their counterparts on the *St. Peter*, they were in desperate need of water but could not find a safe anchorage for the ship. On July 18 they stood a few miles offshore at a depth of seventy fathoms and a latitude, as determined by a series of noon sights, of 57°50' north, while a party led by the fleet master Avraam Dement'ev, along with three sailors, four soldiers, a cannoneer, and two Kamchadal interpreters, took the longboat into shore.

They landed at an uncertain location on either Yakobi or Chichagof Island, identified as either Lisianski Strait, Takanis Bay, or Surge Bay.[60]

The Dement'ev party was armed with guns, sabers, a small copper cannon, and two signal rockets, with instructions to signal as soon as they landed and keep a fire burning at night. They carried a week's provisions, two large water casks to be filled ashore, and trade goods including "a copper and an iron kettle, two hundred beads, three packages of Chinese tobacco, one piece of nankeen, one piece of damask, five rattles, and a paper of needles," with Chirikov contributing ten one-ruble coins to "distribute among the inhabitants as you think best."[61] They were last seen approaching shore at four o'clock in the afternoon on July 18 in fine weather and were never heard from again. The *St. Paul* attempted to maintain position against variable wind and currents but on the twentieth, in thick weather, was driven offshore. On July 23 they returned to the original location, fired their guns, and noticed a fire on the beach. At seven o'clock in the evening they fired another seven signals, observing that "each time we fired the cannon, the fire on shore flared up."[62]

On the twenty-fourth, the remaining boat, a small dory under the command of the boatswain Sidor Savel'ev, who was accompanied by the sailor Dimitri Fadeev, was sent ashore to investigate, carrying tools and materials to repair the longboat as well as the carpenter Nariazhev (or Fedor) Polkovnikov and the caulker Elistrat Gorin. They were instructed to signal upon landing, leave the repair crew, and bring Dement'ev back to the ship. There was a light wind, allowing the ship to sail "quite close to the land, on which a heavy sea was running."[63] At six o'clock the dory was seen approaching shore but failed to signal as instructed and was never heard from again. The *St. Paul* stood by until one o'clock the next afternoon, when two small boats appeared heading toward them from the bay. Assuming these to be the missing crewmen, sail was raised, and all hands were mustered on deck, only to discover that "they were not our boats, because their bows were sharp and the men did not row as we do but paddled."[64]

The smaller canoe, in which four people could be distinguished, approached closer; three were paddling and one was standing at the stern and wearing red. "We saw them stand up, motion with their hands, and heard them call twice, 'Agai, Agai,'" Chirikov reported, "and then they turned about and paddled for the shore."[65]

The outer coast of Yakobi and Chichagof Islands, guarded by a maze of rocks, is hazardous to sailing vessels but hospitable, in fair weather, to small craft. All three conjectured landing sites offer excellent shelter for small boats. Dangers are marked by breakers and kelp; freshwater would have been easy to find; the bays and estuaries were rich in fish, birds, and game; salmon would soon be returning to the streams to spawn; and berries would have been ripe. Nine Russians and two Kamchadals who had been crammed together at sea, with sixty-four others and without adequate food or water, would have passed the breaking outer rocks and entered a placid bay, leaving the *St. Paul* to disappear from sight as they glided along a forested shoreline into a midsummer's afternoon in paradise on earth. Stepping ashore, as ravens sounded the alarm, they would have been unsteady on their feet after six weeks on the rolling decks of the *St. Paul*.

They had landed among the Tlingits, who, distinguished into some eighteen geographic divisions intertwined by a complex network of social and hereditary ties, occupied some eleven thousand miles of coastline and dispersed in summer from their central winter villages to make the circuit of outlying camps. The Tlingits were fierce warriors, sophisticated traders, and expert technicians. They measured their wealth in number of slaves and were accustomed to exchanging hostages in negotiating for safe passage and trading privileges with outside groups. They were skilled at working native copper and familiar with iron that arrived in wreckage drifting from Asia, which they burned to free the metal embedded in the wood. Their matrilineal social structure, based on two persistent moieties and marriage across clans, was robust and independent of scale. In 1741, with the coastal glaciers receding, they were expanding northwestward into skin-boat territory and had not yet suffered the devastating smallpox epidemic that seems to have been introduced by the Spanish visit of 1775.

Yakobi Island and the outer coast of Chichagof Island belonged to the Hoonah tribe, or Kwaan, who controlled the right-of-way by which parties from other divisions traveled north to trade seasonally with Eyak and Chugach from the Gulf of Alaska and Prince William Sound. The French navigator La Pérouse, who visited Lituya Bay at the northwestern boundary of Hoonah territory in July 1786, noted that "every day we saw fresh canoes enter the bay; and every day whole villages departed," estimating that "seven or eight hundred" individuals passed through in just twenty-seven days.[66]

The *St. Paul* had tacked back and forth for three days searching for a place to land. Tlingit observers were surely aware of their approach. It is unlikely that the seaworthy, maneuverable longboat met with natural misadventure in fine weather on a calm summer afternoon and more likely that the Dement'ev party pulled safely in to shore to find themselves facing well-armed Tlingit representatives and a predicament: their instructions were "as soon as you land signal to us with a rocket," but they had also been ordered to "look about for human beings; if you find them, be gentle with them and present them with a few small presents . . . do not offer them any offense."[67]

The Tlingits might have taken the visitors captive, not as an act of hostility, but as a prelude to trade negotiations with those on the larger ship that remained offshore. The landing party in the smaller boat might also have been captured, or swamped by the heavy seas running that day. Chirikov suspected the two canoes that approached the *St. Paul* had been sent to plunder an empty ship and were surprised by the number of Russians who appeared on deck. It is more probable they were sent to invite the visitors to shore.

All we have from the Tlingit side is "Agai! Agai!"—the first Tlingit words to be written down. As the Tlingit historians Nora and Richard Dauenhauer explain, "This spelling could reflect Russians hearing and writing for the first time the Tlingit command 'Axáa!' or 'Ayxáa!' meaning 'Paddle!'"[68] In surviving Tlingit accounts of the first visits by sailing ships, sails are described as the wings of White Raven, with only the bravest daring to look at the sails or approach the ships. Raven, the central actor in Tlingit cosmology, had been white before being turned black when escaping up a chimney after stealing water at the beginning of the world. It was believed that anyone looking directly at White Raven, when White Raven returned, would be turned to stone. The sudden retreat at the unfurling of the *St. Paul*'s sails was followed by what might have been a disastrous misunderstanding on Chirikov's part. "I ordered white kerchiefs to be waved as an invitation for them to board our ship," Chirikov reported, "but it did no good."[69]

The *St. Paul* stood by for the rest of the day, the weather remaining calm. It was now eight days since the longboat had gone ashore. "The action of the natives, their fear to come close to us, made us suspect that they had either killed our men or held them," Chirikov concluded, hanging a lantern in the rigging and standing a safe distance offshore during the night. At eleven o'clock in the morning on the twenty-sixth, they left the area and began sailing slowly northward along the coast, and on July 27, after a ship's council where the decisive argument was the shortage of drinking water and the lack of means to obtain any, it was agreed "to start back at once for the Harbor of St. Peter and St. Paul."[70] The survivors onshore, if any, were left behind.

No conclusive answers to the mystery of the Chirikov landing parties have been found. A petroglyph at Surge Bay, the site of a permanent Tlingit settlement in the eighteenth century, shows what appears to be a small vessel with four oars on the port side, three people at the stern, one of them steering, and two rectangular objects that have been interpreted as masts or sails, but seem to me to be the artist's representation of the two large water barrels that were carried in the boat. Four oarsmen on each side totals eight, plus the three people at the stern adds up to eleven, the exact complement with which Dement'ev went ashore.

In Tlingit oral history, the Russians deserted by choice. "They took this opportunity to escape the cruel and harsh conditions on the Russian ship," says Mark Jacobs Jr. of Sitka. "As they left the ship, they decided among themselves that they would eventually perish in the hazardous waters of the North Pacific. Why suffer under a cruel command until then? The decision to desert was easy. They were accepted and treated with respect, instead of being murdered as the Russian history tells it. They married Tlingit women and all was well . . . The water kegs, used to replenish the fresh water supply on the ship, were a highly treasured prize, they were converted to store seal oil . . . The boat was burned and the nails salvaged."[71]

The *St. Paul* returned safely to Kamchatka, losing only six men to scurvy on the return voyage, plus La Croyère, who died after they dropped anchor in Avacha Bay on October 9. By August 1 they were reduced to boiled rye mush once a day, then once every other day, and, after September 14, once a week, subsisting otherwise on cold biscuits, salt beef soaked in seawater, and a daily allowance of vodka that was increased by two cups as water ran out. They distilled seawater and collected rainwater from the sails.

On September 9, off Adak Island, they were visited by seven Aleut kayakers, whom they asked to fill a barrel with water from shore. "They understood what we meant, but they would not take the barrel and showed us that they had bladders for that purpose," noted Chirikov. "Three of them paddled towards the beach and returned with water. When they came alongside one of them held up a bladder and indicated that he wished to have a knife in payment. This was given to him, but instead of handing over the bladder, he passed it to the second man, who also demanded a knife. When he got it he passed the bladder to the third man, who equally insisted on a knife." This "proves that their conscience is not highly developed," Chirikov added, after paying the asking price.[72]

The knives the visitors exchanged for the water that saved their lives were not that different from the knives made of native copper or salvaged iron that were already in circulation on the Northwest Coast. The vessels they arrived in were larger, but less maneuverable, than the indigenous skin boats and dugout canoes. Even the Russian firearms were less effective than the atlatl (throwing stick) and spear for hunting or combat on the open ocean in small boats. The Russian conscience, or intellect, was no more developed than that of the Tlingits or Aleuts. Only in their alphabet and written language did they have history on their side.

LAST OF THE APACHES

The first large-scale, high-speed wireless data communications network in North America commenced operations on May 1, 1886. The all-optical network, encompassing twenty-seven stations extending across sixty thousand square miles of New Mexico and Arizona, carried 2,264 messages, averaging fifty words each, over the next five months.

These messages, relayed in Morse code over beams of sunlight between stations as much as fifty miles apart, were only a preliminary flicker of Leibniz's vision, with no further logical processing of the data that were sent, yet signaled a transition as profound as when the Russians had arrived with a written alphabet in 1741.

The target of the campaign was the next-to-last remnant of free-ranging Apache "Indians," amounting to nineteen men, thirteen women, and six children, led by the warriors Naiche and Geronimo, who surrendered to

Pinnacle Pt

View of the Land when Pinnacle Point bore N. ½ W. 6 leas. dist.

the U.S. Army general Nelson Appleton Miles on September 4, 1886. Miles was the self-proclaimed architect of both the new data network and a series of brutal Indian Wars that were drawing to a close. "The Indians are practically a doomed race," he reported in 1891, "and none realize it better than themselves."[1]

Geronimo's small band of renegades, along with more than four hundred of their nonbelligerent reservation-abiding relatives, including the Apache scouts who had assisted the U.S. Army in trailing the elusive fugitives, were declared prisoners of war and exiled to Florida. In violation of the agreement with the scouts, the treaties establishing the reservations, and the promises made to Geronimo and Naiche, those who survived malaria, tuberculosis, homesickness, and the removal of their children to residential schools waited twenty-seven years to return to Arizona and never regained their homes. "I looked in vain for General Miles to send me to that land of which he had spoken," Geronimo mourned in exile. "I longed in vain for the implements, house, and stock that General Miles had promised me."[2]

Four years earlier, the Apaches had fought their last open engagement with the regular forces of the U.S. Army, a battle that took place in Chevelon's Fork of the Cañon Diablo in Arizona on July 17, 1882. The party of White Mountain Apaches, led by the warrior Na-tio-tish and numbering fifty-four fighters, had killed the chief of Indian police and three of his Native scouts at the San Carlos reservation before making their escape into the Tonto basin with fourteen troops of U.S. Cavalry, as well as a posse of civilians who had spilled out of the saloons in Globe, Arizona, in hot pursuit.

The Globe Rangers soon lost the trail as well as their horses, spirited away by the Apaches during the night. The cavalry stayed close on the heels of the escapees, who disappeared into one of the canyons that led through the Mogollon escarpment toward Navajo territory to the north. As Lieutenant Britton Davis, then twenty-two years old, described it, "Across this mesa ran a gigantic slash in the face of the earth, a volcanic crack, some seven hundred yards across and about one thousand feet deep, with almost perpendicular walls for miles on either side of the very steep trail."[3]

Na-tio-tish misjudged the strength of his pursuers, and the intended ambush failed, with the Apaches trapped by five troops of soldiers who proceeded to outflank them from both sides. According to Davis, only two were killed on the government side, while twenty-one Apaches were killed outright and five died later of their wounds. Davis, who arrived in the early morning of the eighteenth, missed the firefight but helped to disarm an Apache woman of eighteen or nineteen, who, with one leg shattered by a rifle ball and a six-month-old infant in her arms, had fired upon the soldiers from a rocky vantage point at daybreak with the last three cartridges she had left. Her leg was amputated, without anesthetic, above the knee, after which she spent a full week, with her infant, limping back to Fort Apache on a government mule.

Second Lieutenant Thomas Cruse, of the E Troop of the Sixth Cavalry, who was known as "Nantan Greenhorn," according to Geronimo's nephew Asa "Ace" Daklugie, "because of his ignorance of the Apaches," was awarded the Medal of Honor for his role in the engagement.[4] He helped lead the advance guard who crossed the canyon at three o'clock in the afternoon to engage the warriors who were lying in wait on the other side. "Bright sunlight was on us until we got to the cañon, scrambling down the precipitous wall to the beautiful stream flowing along its floor," he reported. "Then someone pointed upward and we stared—at stars plain to be seen in midafternoon!"[5]

The Arizona sky, revealing daytime stars that the scattering of sunlight by atmospheric moisture otherwise conceals, transmitted that sunlight with so little attenuation that temperatures reached into the 120s and guns left out in the open became too hot to pick up with bare hands. General Miles enlisted this unfiltered radiation in his campaign to drive the Apaches out of their stronghold in the American Southwest. "I had it in my mind to utilize for our benefit and their discomfiture, the very elements that had been the greatest obstacles in that whole country to their subjugation, namely, the high mountain ranges, the glaring, burning sunlight, and an atmosphere void of moisture," he explained. "As to their being able to signal by the use of fire and smoke and the flashes of some bright piece of metal for a short distance, I thought we could not only equal, but far surpass them in a short time."[6]

Nine years earlier, Miles had led the midwinter campaign against the Nez Percés that ended in the surrender of Chief Joseph and the death of Chief Looking Glass, who inherited both his name and a reflecting necklace from his father, a war leader who "carried a looking-glass, used to direct military manoevres in battles by means of reflected rays of light."[7] Apache warriors also carried small mirrors for signaling. That they were awed or mystified by the telecommunications network deployed against them was a fiction invented by General Miles. Geronimo and Naiche were induced to surrender by false promises, not by messages conveyed by coded beams of light.

Miles established heliographic signal stations at strategic locations across New Mexico and Arizona, relaying observations and commands among his scattered troops. "The messages, flashed by mirrors from peak to peak of the mountains, disheartened the Indians as they crept stealthily or rode swiftly through the valleys, assuring them that all their arts and craft had not availed to conceal their trails," George W. Baird reported, in recounting Miles's exploits for *The Century Magazine*.[8] The renegades, hounded but not defeated, were driven out of Arizona and New Mexico and into the Sierra Madre in Mexico, beyond the heliograph network's reach.

"When they saw heliographic communications flashing across every mountain range, Geronimo and others sent word to Natchez [Naiche] that he had better come in at once and surrender," Miles reported to the secretary of war at the close of the campaign, neglecting to explain that Geronimo and Naiche had laid down their arms only upon his assurance that they would be given safe passage back to the United States, reunited with their families within five days of surrender, returned home after a two-year imprisonment in Florida, and absolved of all criminal charges that the civilian authorities in Arizona were holding over their heads.[9]

Despite being outnumbered by more than one hundred to one, during the entirety of General Miles's pursuit of Geronimo's band of guerrilla warriors, no Apaches were killed. None were taken prisoner until after their surrender, well armed and on their own terms. The "capture" of Geronimo is a myth.

A loose association of autonomous, seminomadic groups united by a common language, strategic intermarriage, and, with the arrival of Spanish conquistadores in the sixteenth century, a common enemy, the Apaches, of Athapaskan descent, had arrived from the north. Their ancestors, after making their way to America from Asia more than fifteen thousand years before, had settled down in the subarctic and Pacific Northwest while those who would become known as Apaches had kept pushing south. They were the Dineh, or People, distinguished into a spectrum of subpopulations whose fluidity was its strength. Like desert cacti that propagate by growing small limbs that break off to spawn new cacti. the Apaches were expert at spinning off small autonomous bands.

The Apaches were divided into seven major tribes. The Western Apaches, to consider just one of the seven divisions, were further divided into five distinct subtribal groups, more than a dozen geographically defined bands, and some sixty-two kinship-defined clans. In contemporary popular literature, "Apaches" referred to whatever group happened to be at odds with the authorities at the time. They practiced only rudimentary agriculture, with most of their diet obtained from indigenous plants, raided livestock, and wild game. In the arid Southwest, meat dried before it decayed, allowing them to traverse time and distance on field-dehydrated rations from one kill to the next. They resisted the presumption that beef cattle, raised by settlers who had driven off the existing game animals, were not fair game themselves.

The name Apache seems to have originated from their designation as outsider or enemy (*apachu*) by the Zuni, who were settled agrarians whose livestock and stores of grain were the subject of predatory raids. The characteristic Southwest Pueblo culture, with fortified settlements that dominate the archaeological record, was the visible manifestation of the Apaches who roamed the surrounding landscape, forcing the construction of elaborate defenses but otherwise leaving little trace. Apache territory extended deep into the mountains of Chihuahua and Sonora in present-day Mexico: the last refuge for those such as Geronimo and Naiche who refused to submit to captivity in the United States.

The first recorded European visitor to Apache territory was a black, Arabic-speaking, Muslim turned Christian slave from Morocco named Esteban (or Estevanico) de Dorantes, a survivor of the ill-fated Narváez expedition that had landed, with six hundred men and eighty horses, on the west coast of Florida in April 1528. When the survivors of the three hundred men and forty horses who had ventured to explore inland returned to the coast, they found that the remainder of the party had departed with the ships and abandoned them to their fate. Disease, attacks by natives, shipwreck of the crude vessels they built to try to make their way westward past the mouth of the Mississippi, and starvation to the point of cannibalism killed all but two who were reported to have survived as captives, and four who escaped captivity and made their way to the Pacific coast.

Esteban; his master, Andrés Dorantes; and two companions, Álvar Núñez Cabeza de Vaca and Alonso del Castillo Maldonado, were captured and enslaved for five years on the Texas coast. After making their escape, they styled themselves as faith healers and wandering prophets, gaining a wide following and passing from one tribe to another until, almost two years later, they reached the Gulf of California, where they met up with a syndicate of slave hunters from Sinaloa who were depopulating the local settlements and sought to add the castaways' entourage to the hundreds of prisoners they already had in chains. As recounted by Cabeza de Vaca, their Native companions refused to believe that they belonged to the same race as the slavers, "for we had come from sunrise, while the others came from where the sun sets; that we cured the sick, while the others killed those who were healthy; that we went naked and shoeless, while the others wore clothes and went on horseback."[10] The four survivors arrived in Mexico City in July 1536. They were provided with clothing by Governor Nuño de Guzmán, "but for many days," in Cabeza de Vaca's account, "I could bear no clothing, nor could we sleep, except on the bare floor."[11]

Esteban's freedom was purchased in 1537 by Antonio de Mendoza, the viceroy of New Spain, who dispatched him on a reconnaissance mission to the fabled Seven Cities of Cíbola with the French-born friar Marcos de Niza in 1539. Marcos, who had walked barefoot from Peru, returned with a much-embellished description of the riches to be found among the pueblos to the north—and without Esteban, who was the sole member of the Marcos party to reach the Seven Cities, but only as a dismembered corpse.

According to Pedro de Castañeda de Nájera, a member of the Coronado expedition who did not join the journey to Cíbola but later compiled an account, Esteban had occupied himself, during the slow start to the venture, by "taking the women the Indians gave him, collecting turquoises, and amassing a quantity of both." The friars, while expressing disapproval at this behavior, nonetheless delegated Esteban to lead the way to Cíbola, so that when "they arrived they would have nothing to think about other than collecting reports." Esteban passed unharmed through Apache territory, arriving at the Zuni fortifications "loaded with numerous turquoises and some beautiful women," where he was sequestered outside the walls of the settlement and asked to explain himself. After three full days of deliberation, the Zuni council decided to execute him, "because it seemed nonsense to them to say that the land he was coming from was one of white people who had sent him, when he was black."[12]

A year later, following up on Friar Marcos's report, an armed party under Francisco Vázquez de Coronado, with seventy-five men, set out northward on June 24, 1540, from Chichilticale, an abandoned settlement in Chiricahua Apache territory in the extreme southeast corner of present-day Arizona, in the Sulphur Springs Valley near Apache Pass. They described Chichilticale, once an outpost of the Cíbola settlements, as a "fortress [that] must have been destroyed by the people of the district, who are the most barbarous people that have yet been seen."[13] These nomads, not yet designated Apaches, and with no fixed assets of interest to the Spaniards, survived by hunting and lived in scattered camps. From here northward to the Zuni pueblos the Spaniards described the country as "unsettled" (*despoblado*), although unsettled does not mean uninhabited. The expedition was following a well-worn trading path.

Coronado, in contrast to Esteban, found the landscape inhospitable and lost many horses and men. The expedition arrived, nearing starvation, at the gates of the Zuni settlement, where a standoff ensued, with a proclamation in the name of His Majesty the king of Spain answered by a single Zuni arrow that "pierced the gown of Friar Luis . . . which, blessed be God, did him no harm."[14] Upon further evidence of Zuni hostility, the friars, with hunger winning out over their instructions to pursue diplomacy, gave the approval to attack.

The Zuni responded with a hail of arrows, followed by stones hurled from the rooftops, which, according to Coronado, were directed "against me because my armor was gilded and glittered, and on this account I was hurt more than the rest." Coronado was knocked unconscious and carried to safety during the battle, surviving his injuries although left "somewhat sore from the stones." Returning fire with crossbows and harquebuses, the Spaniards prevailed, driving the defenders out of the pueblo and settling in to feast on the stores the Zunis had left behind. They found "the best corn cakes I have seen anywhere," according to Coronado, along with wild turkeys that, although kept by the Zunis "merely for the sake of procuring the feathers," were found to be "very good, and better than those of Mexico." One of Coronado's officers described the plunder as "what we needed more than gold and silver, and that was much corn and beans and fowls, and salt, the best and whitest that I have seen in all my life."[15]

The Apache inhabitants of *los despoblados* kept the invaders under surveillance but melted away to let them pass. "No Indians were seen during the first day's march," Coronado reported to the viceroy, "after which four Indians came out with signs of peace, saying that they had been sent to that desert place to say that we were welcome, and that on the next day the tribe would provide the whole force with food."[16] The Spaniards suspected the Apaches of preparing an ambush in one of the passes, but once discovered and challenged, the Apaches fled.

Coronado and his officers marveled at the Apache use of smoke signals, "which were answered from a distance with as much coordination as we would have known how to do ourselves."[17] To what extent the Apaches and the Zunis coordinated in tracking and opposing the Coronado expedition is unknown, but the Zunis later explained to Coronado that the execution of Esteban was carried out "because the Indians of Chichilticale said that he was a bad man, and [had] assaulted their women, whom the Indians love better than themselves."[18] Esteban's bones were distributed among the settlements as confirmation that the interloper was both mortal and now dead. The Zunis also sent a messenger back to Chichilticale advising the residents to kill any further intruders and, if they were unable to do so, to request assistance from the Zunis, who "would come and do it right."[19]

The Chiricahuas were thus the first Apaches to meet the "White-Eyes," as they termed the foreigners, and, 346 years later, the last to give up the fight. They entered into an uneasy coexistence with the Spaniards and then the Mexicans, who sequestered themselves in fortified towns while the Apaches retained control of the backcountry, taking advantage of the colonial outposts and their easily ambushed supply trains as a source of valuables such as firearms, textiles, and the potent distilled liquor mescal. The heart of Chiricahua territory was never permanently occupied, and for three hundred years all attempts at settlement were driven out. Unable to defeat the Apaches in battle, the Mexican government, beginning in 1835, offered a bounty on Apache scalps: twenty-five pesos in silver for children, fifty pesos for females, and one hundred pesos for males. Many Apaches as well as non-Apaches were killed for their scalps.

In the aftermath of the Mexican-American War of 1846–1848 and the Gadsden Purchase of 1854, most of the traditional Apache territory was conveyed to the United States. The new boundary remained permeable: not only to the Apaches who raided across the border in both directions but also to freelance American bounty hunters who collected payment in Mexican silver for "Apache" scalps and to freelance Mexican slave traders who ranged across the border into the United States.

In addition to these Mexicans from the south, a growing flood of settlers, prospectors, military expeditions, and camp followers, along with outlaws and drifters driven out of gold-rush-era California, began pushing into Apache territory from the west. These new arrivals cut their hair short and wore their beards long, contrary to the sensibilities of the Apaches, whose warriors wore their hair shoulder length with all traces of facial hair plucked out.

The Americans appeared to offer an improvement over the Mexicans and were welcomed, at first. As the settlers appropriated water sources, minerals, and rangelands, the Apaches had second thoughts and began pushing back. Gangs of frontiersmen, supported by regular soldiers, retaliated against any resistance, with atrocities escalating on both sides. Regular and irregular forces destroyed not only the Apache encampments but also their food supplies, forcing the survivors to become even more aggressive, or starve. Many settlers were killed, and when U.S. soldiers began withdrawing to fight in the Civil War, it looked to the Apaches as if they were driving the invaders back.

Arizona and New Mexico divided their allegiance between the Union and the Confederate camps. The Apaches suffered from both sides. "Use all means to persuade the Apaches or any other tribe to come in for the purpose of making peace, and when you get them together kill all the grown Indians and take the children prisoners and sell them to defray the expense," Lieutenant Colonel John R. Baylor, Confederate governor of Arizona, instructed the commander of the Arizona guards.[20] Jefferson Davis and his secretary of war, G. W. Randolph, rescinded the order and relieved Baylor of his command, while Brigadier General H. H. Sibley, in command of the Confederate army in New Mexico, sought "to encourage private enterprises" against the Navajos and Apaches by instituting policies "to legalize the enslaving of them."[21] On the Union side, General James H. Carleton, after taking command of the Department of New Mexico, ordered Colonel Kit Carson to "make war upon the Mescaleros and upon all other Indians you may find in the Mescalero country, until further orders. All Indian men of that tribe are to be killed whenever and wherever you can find them. The women and children will not be harmed, but you will take them prisoners."[22]

Geronimo was elevated to war leader following the extrajudicial execution of Mangas Coloradas, chief of the Mimbreños, by U.S. soldiers in 1863. Mangas was born in the early 1790s, grew up at war with the Mexicans, and at first sought to cooperate with the Americans, until, after reportedly having been captured and flogged, he turned against them and formed an alliance with the Chiricahuas and their leader, Cochise. The combined forces held their ground until Carleton's party of California Volunteers, carrying howitzers, used these "wagon guns" to shell Apache defensive positions in Apache Pass. Mangas then decided, against the advice of the other warriors, to parley under a flag of truce with a group of settlers and prospectors led by Joseph Reddeford Walker and accompanied by a group of cavalry under Edmond D. Shirland, also of the California Volunteers.

The party was camped near the abandoned mining town of Pinos Altos on January 19, 1863, when, as recounted by Daniel Ellis Conner, they "succeeded by hoisting the white flag of drawing some Indians into our sight." With the Apaches within range, Walker's men "presented their guns and ordered Mangas to stand still and surrender."[23] Mangas was taken to Fort McLane and delivered into the custody of General Joseph R. West, who placed him under the guard of two sentries who were given instructions, according to the official record, to shoot the captive if he made any attempt to escape.

"It was a bitter cold on that bleak prairie," reported Conner. "No fire was kept burning except the one at the junction of the two guard beats, where Mangas lay . . . A while before midnight I noticed Mangas moving now and then . . . I could see them plainly by the firelight as they were engaged in heating their fixed bayonets in the fire and putting them to the feet and naked legs of Mangas, who was from time to time trying to shield his limbs from the hot steel." Just before midnight, Conner, watching from a distance, saw that "Mangas raised himself upon his left elbow and began to expostulate in a vigorous way by telling the sentinels in Spanish that he was no child to be playing with." The guards took their cue. "He had hardly begun his exclamation when both sentinels promptly brought down their Minie muskets to bear on him and fired, nearly at the same time."[24]

At daybreak, the body was scalped by a soldier from California named John T. Wright before being buried in a shallow grave. A few days later, the corpse was exhumed and decapitated, and the head was boiled in a large iron pot. General West filed a report stating that Mangas had been killed after attempting to escape, while the army surgeon D. B. Sturgeon delivered the skull to the phrenologist Orson S. Fowler, who reported that it was the "broadest human skull I have ever seen" and beyond "comparison . . . as to Cunning, Destruction, or the perceptives." He added that Dr. Sturgeon "saw this Indian a few minutes after he was shot, and prepared this skull expressly for me, so that its identity is thus assured."[25] The skull was put on display in New York.

With the end of the Civil War, official policy shifted from extermination to concentration. The army was deployed to drive the remaining Apaches onto four designated reservations, where civilian agents and contractors were assigned responsibility for their welfare but were all too often preoccupied with enriching themselves. The tragic history of the Apaches is as much the story of the U.S. Army (under the secretary of war) versus the Bureau of Indian Affairs (under the secretary of the interior) as it is the story of the U.S. Army against the Apaches themselves. The local Indian agencies, riddled with corruption, often treated those on the reservations no better than did the professional soldiers and scouts assigned to kill or recapture them when they escaped.

The Apaches, who had gained little from three hundred years of Spanish occupation except the horse, gained little from the Americans except the Winchester repeating rifle, "the gun that won the West." When breech-loading rifles came into circulation after the Civil War, the Apaches equipped themselves, as Britton Davis noted, "with the latest models of Winchester magazine rifles, a better arm than the single shot Springfield with which our soldiers and scouts were armed."[26] When surprise was necessary or ammunition ran out, or to not draw fire at night, Apache warriors reverted to bows, arrows, and lances. The lance was a more effective weapon than the army saber in close quarters, because it could reach farther and was easy to withdraw. "We liked long sharp points on our lances because they pulled out easily," Daklugie explained. Attacks with bows and arrows were feared by the Arizona settlers more than Apaches attacking with guns. "I knew men who could put seven arrows in the air before the first one that was fired fell to the earth," remembered Daklugie. "There was a time that I could do that myself."[27]

The Apaches resisted both the initial confinement to reservations and a series of relocations when land that had been granted "in perpetuity" was reclaimed. Between 1873 and 1877, those who had settled at Warm Springs and other agencies were driven in a series of forced marches to the inhospitable San Carlos reservation, where they were consolidated with their long-standing enemies under a system that appeared designed to spawn discontent. "San Carlos!" Daklugie remembered. "That was the worst place in all the great territory stolen from the Apaches. If anybody had ever lived there permanently, no Apache knew of it. Where there is no grass there is no game . . . The heat was terrible. The insects were terrible. The water was terrible."[28]

The captives were kept on subsistence rations and forced to wear numbered identity tags around their necks. "From the time that we, as peaceful noncombatants, were driven from our reservation until we were herded aboard a train at Holbrook, Arizona, and shipped to Florida in 1886, we had been hunted through the forests and plains of our own land as though we were wild animals," James Kaywaykla, then a young boy and later Naiche's son-in-law, explained. "Until I was about ten years old I did not know that people died except by violence."[29] It was no accident that they had been confined to a reservation where the water was bad, crops failed to thrive, and game was scarce. "The Apache was a hard foe to subdue," according to John Gregory Bourke, "not because he was full of wiles and tricks and experienced in all that pertains to the art of war, but because he had so few artificial wants and depended almost absolutely on what his great mother—Nature—stood ready to supply. Our government had never been able to starve any of them until it had them placed on a reservation."[30]

Bourke was one of a succession of army officers who were swayed by the courage and integrity of the Apaches into speaking up in their defense. "In treachery, broken pledges on the part of high officials, lies, thievery, slaughter of defenseless women and children, and every crime in the catalogue of man's inhumanity to man the Indian was a mere amateur," Britton Davis, who left the army in frustration, reminisced. "His crimes were retail, ours wholesale. We learned his methods in that line and with our superior intelligence, improved upon them. The only thing we did not adopt was his method of torturing prisoners. We had better ones of our own."[31]

Bourke served as adjutant to General George Crook, whom he described as "more Indian than the Indians!" Crook's style was opposite to that of General Miles. Miles's autobiography, *Serving the Republic*, was a self-serving 340-page chronicle, bound in blue cloth emblazoned in gold leaf with a facsimile of his own uniform's braided stars. Crook, who submitted terse official reports and left his own autobiography unfinished at

82270989104473536

303.48305 DYSON

his death, dressed in a weather-beaten canvas coat and wore a trademark khaki helmet for a hat. Although commander of the cavalry, he rode a mule named Apache who carried him through all his campaigns and upon Crook's death in 1890 was "protected and fed in honorable retirement" in Omaha, Nebraska, headquarters of the Department of the Platte.

Crook's success derived as much from his preference for mules over horses as it did from his familiarity with the enemy and his recruitment of Apache scouts. He reformed the methods by which the recalcitrant animals were loaded and managed, allowing horses to be left behind. At least once the brigadier general was mistaken for a mule packer himself. "Regular troops are as helpless as a whale attacked by a school of swordfish," he explained to the secretary of war.[32] The only way to track down renegade Apaches was to employ cooperative Apaches as scouts. "There has never been any success in operations against these Indians, unless Indian scouts were used either as auxiliaries or independent of other support," he noted in the official summary of his campaign.[33]

He exploited long-standing rivalries between different bands and enlisted a series of otherwise misfit individuals, drawn from the rough edges of the frontier, as intermediaries between the soldiers and the scouts. Al Sieber, a German by birth who killed more Apaches than anyone else on record, served seventeen years as chief of scouts and was wounded twenty-nine times—or thirty, if you count being crushed to death in 1907 by a large boulder that rolled over him, under suspicious circumstances, while supervising a crew of Apache laborers building a road to the Roosevelt Dam. Mickey Free, whose mother was Mexican and father Irish, had been captured and raised by Apaches as a child and served as a trilingual interpreter (or misinterpreter, according to Geronimo) as well as a tracker and hired gun. Tom Horn was a sharpshooter and mule packer with a checkered past who found his services to Crook's campaigns and the U.S. Army insufficient, in the end, to keep him from being hanged. "My mother was a tall, powerful woman," Horn remembered while awaiting his appointment with the gallows, "and she would whip me and cry, and tell me how much good she was trying to do me by breaking me of my Indian ways."[34]

In 1882, Crook was reassigned to the San Carlos reservation in the aftermath of an uprising precipitated by the unprovoked killing of Noche-del-klinne, a reclusive Apache prophet possessed by a vision of a resurrection of the dead Apache warriors and a restoration of the Apache homeland and way of life. Noche-del-klinne, who was accused by the army of fomenting rebellion, was, to the contrary, teaching that by taking the path of non-violence, the Apaches would survive to outlive the invading civilization that would ultimately destroy itself.

Crook spent his entire first month living in a tent a full mile from the army post, gathering firsthand testimony from the Apaches who had been involved in the fight. Only then did he invite his own officers out to his camp to give their side of the story and proceed to reconstruct relations from there. "When you left here, everything was in good shape; there were no bad Indians out," testified one of his Apache informants, Alchisay. "The officers you had here were all taken away and new ones came in—a different kind. The good ones must have all been taken away, and the bad ones sent in their places. We couldn't make out what they wanted: one day they seemed to want one thing: the next day, something else."[35]

Under Crook's administration, between 1882 and the outbreak led by Geronimo in 1885, Apaches who agreed to wear numbered identity tags were allowed to live in outlying areas of the reservation, rather than being reduced to captive subsistence within sight of the military forts. "Every Indian who can't produce such a tag shall be considered a hostile," he warned. "I'll count you every day until all the Chiricahuas and others now out have come in."[36] Crook exposed a network of corrupt Indian agents and subcontractors, installing replacements like Britton Davis, to whom he assigned the relocated Chiricahua settlement at Turkey Creek, and Charles Gatewood, who gained the name Long-nose along with Geronimo's personal trust. In 1883 the White Mountain Apaches raised 135,000 pounds of potatoes, 180,000 pounds of beans, 200,000 pounds of barley, and 2,625,000 pounds of corn.[37]

Davis, assigned to distribute heavy steel plows and other agricultural implements issued by the government, laughed alongside his charges as Geronimo's warriors put their light war ponies who "had never even been broken to saddle" into harnesses "big enough for two of them" and went galloping across the San Carlos River bottom, the plow blades skating across the hardpan, leaving barely a scratch. "My feelings toward them began to change," Davis reported after a year alone among his former adversaries. "That ill-defined impression that they were something a little better than animals but not quite human; something to be on your guard against; something to be eternally watched with suspicion and killed with no more compunction than one would kill a coyote; the feeling that there could be no possible ground upon which we could meet as man to man, passed away."[38]

When trouble broke out in May 1885, the uprising was once again among the Chiricahuas, whose numbers had dwindled to fewer than five hundred individuals, among them a small band of warriors that included the direct descendants of Victorio, Mangas Coloradas, and Cochise. Naiche, a grandson of Mangas Coloradas and youngest son of Cochise, was their hereditary chief. Geronimo was their strategist and spokesman, being Chiricahua by marriage, not by birth. "I was born in No-doyohn Canyon, Arizona, June, 1829 . . . around the headwaters of the Gila River," Geronimo remembered in 1906. "I was warmed by the sun, rocked by the winds, and sheltered by the trees."[39] His Apache name was Goyaałé: "he who yawns."

At age seventeen Geronimo married his childhood sweetheart, Alope, whose father asked a "herd of ponies" for her. Geronimo returned with the requisite number of ponies in only a few days. The "fair Alope" was an artist, according to her new husband, and "drew many pictures on the walls of our home." In a short time, "three children came to us—children that played, loitered, and worked as I had done."[40] Life was idyllic until the spring of 1858, when, on a trading expedition to Sonora, the band's encampment, with the warriors absent, was attacked by Mexican troops.

Geronimo's mother, along with Alope and the three small children, was among the dead.

Geronimo joined the subsequent revenge expedition under the leadership of Cochise. Driven by a thirst for retribution and evidencing powers that would have distinguished him as a shaman in more peaceful times, he outperformed all the other warriors, even the legendary Cochise, and soon began leading war parties against the Mexicans himself. Legend has it that as he was killing Mexican soldiers, they appealed for assistance to Saint Jerome (Hieronymos). The appeal failed, but the name stuck.

A natural statesman, Geronimo had a keen sense of his own role in history, driven in part by premonitions as to how the tragedy would play out. In 1905, he was granted leave from his imprisonment to ride in Theodore Roosevelt's inaugural parade, although a personal appeal to the new president to allow his people to return home from exile was rebuffed. He had nine wives, led four major outbreaks from the reservations, and was only captured against his will once. As a high-value target, he could advance someone's career by offering to surrender or scuttle someone's career by making his escape. He believed that both General Miles and General Crook had betrayed him and regretted having surrendered, wishing he had followed the example of Victorio, who had fought to the last bullet and then killed himself with his own knife.

Geronimo never apologized and maintained that his later killings, most committed in Mexico, were executed in self-defense. "We felt that every man's hand was against us. If we returned to the reservation we would be put in prison and killed; if we stayed in Mexico they would continue to send soldiers to fight us; so we gave no quarter to anyone."[41] Naiche, interviewed by General Crook in 1890, agreed: "We were afraid. It was war. Anybody who saw us would kill us, and we did the same thing. We had to if we wanted to live."[42]

Naiche and Geronimo, along with thirty-two other warriors, eight boys old enough to carry weapons, and ninety-two women and children, left the

San Carlos reservation on May 18, 1885, cutting the telegraph line as they made their escape, and covering 120 miles before stopping to rest. Crook had no trouble raising a party of scouts to go in pursuit. They "were excited and happy over the prospect of going out on another campaign," Jason Betzinez, one of Geronimo's cousins, remembered. "Hardship and danger meant nothing, adventure was what they wanted and it didn't seem to matter that they were going to fight their own people."[43]

The fugitives left a trail of casualties on their way to Mexico and in subsequent incursions back into the United States, including two young brothers, Martin and James McKinn, who were surprised while herding their father's cattle along Gallina Creek, a tributary of the Mimbres River, in New Mexico, on September 11, 1885. Martin, fifteen, was killed by a single rifle shot, while James, eleven, was taken captive after "putting up quite a battle with sticks and rocks."[44]

Crook's forces, including more than two hundred scouts, followed the trail into the Sierra Madre, but Geronimo and Naiche could not be cornered or otherwise induced to fight. In January 1886, from a stronghold with multiple escape routes, Geronimo and Naiche sent two of their wives as emissaries and agreed to meet with Crook, at Cañon de los Embudos, Mexico, a conference site selected by the Apaches, just below the Arizona border, in two months.

Crook waited for the Apaches, who were settling scores with the Mexicans for the last time. Geronimo feared the Mexicans, who had a price on his head, and the citizens of Tucson, who hoped to see him hanged, more than he feared the U.S. soldiers, whom he was counting on to provide safe passage back to the United States. "I proceeded to the point where the Indians were in camp, and on March 25th, 1886, had my first interview with them," Crook reported. "I found the hostiles, though tired of the constant hounding of the campaign, in superb physical condition, armed to the teeth, and with an abundance of ammunition."[45]

John Bourke noticed a young boy in the Apache camp with "blue-gray eyes, badly freckled, light eyebrows and lashes, much tanned and blistered by the sun," who "wore an old and once-white handkerchief on his head which covered it so tightly that the hair could not be seen." This was James McKinn, who now went by the name Santiago, given to him by Chihua-hua, to whom he had become attached. "He seemed to be kindly treated by his young companions," Bourke noted, "and there was no interference with our talk."[46]

McKinn was photographed by the Tombstone photographer Camillus S. Fly and described by Charles Lummis, covering the campaign for the *Los Angeles Times*. "Wildest in the rough sports of the bronco boys was one figure which you would single out at a glance," Lummis reported. "This poor child, scaly with dirt, wild as a coyote, made my eyes a bit damp." McKinn, now "Apachefied" and "absolutely Indianized," spoke only in Apache and "utterly refused to budge toward the fort with any white man, and Chihuahua had to bring him up." Told that he was to be taken back to his father and mother, McKinn answered in Apache "that he didn't want to go back—he wanted to always stay with the Indians. All sorts of rosy pictures of the delights of home were drawn, but he would [have] none of them, and acted like a young wild animal in a trap."[47]

That this ragtag band could hold out against the U.S. Army was em-barrassing to the government, but the spectacle of their negotiating from a position of strength was worse. President Grover Cleveland continued to insist on unconditional surrender while authorizing Crook to assure the Apaches that they would be treated as prisoners of war and their lives would be spared. Crook was ordered, with intentional ambiguity, "to make no promises at all to the hostiles, unless it is necessary to secure the surren-der."[48] The president's ultimatum, delivered by Crook, was met with a firm refusal from Geronimo and Naiche, who preferred to fight to the death.

After two days of discussion, a compromise was reached. On March 27, 1886, Crook accepted Geronimo and Naiche's surrender, under the condi-tion that "they should be sent east for not exceeding two years, taking with them such of their families as so desired."[49] Geronimo, who had opened the negotiations on the twenty-fifth with a lengthy chronicle of the injustices against the Apaches, spoke very few words on the twenty-seventh as they were closed. "Once I moved about like the wind," he said. "Now I surrender to you and that is all."[50]

President Cleveland had second thoughts. "The President cannot assent to the surrender of the hostiles on the terms of their imprisonment East for two years with the understanding of their return to the reservation," General Philip H. Sheridan telegraphed Crook on March 30. Unless the surrender was unconditional, "you must make at once such disposition of your troops as will insure against further hostilities by completing the destruction of the hostiles," he ordered.[51] By the time Crook received these instructions to betray the agreement, Geronimo and Naiche had made their escape.

On the evening of March 27, with the Apaches and the troops camped next to each other just south of the border for the night, a Swiss American bootlegger and dealer in stolen livestock named Robert Tribolet supplied the Apache encampment with several gallons of cheap whiskey, accompanied by a warning that Geronimo and his accomplices were to be hanged if they returned to Arizona and would be better off taking flight.

This misinformation was presented less for the benefit of Geronimo and more for the benefit of the syndicate of Tucson merchants who were profiting from the government contracts required to support the prolonged anti-Apache campaign. They needed a renegade, and Geronimo was the last one left. Also, considerable cattle rustling by non-Indian malefactors was being attributed to the Apaches, and it would be bad for Tribolet's business if no outlaw Apaches were at large to take the blame.

"Alchise and Kaetenna came and awakened General Crook before it was yet daylight of March 28th and informed him that Nachita, one of the Chiricahua chiefs, was so drunk he couldn't stand up and was lying prone on the ground," reported Bourke, who accompanied Crook in trying to defuse the situation before anyone got hurt. "Pretty soon we came upon Geronimo, Kuthli, and three other Chiricahua warriors riding on two mules, all drunk as lords."[52]

The intervention came too late. Crook would later ask Naiche why they had taken flight. "I thought all who were taken away would die," Naiche answered. As to why they got drunk before leaving, "there was a lot of whiskey there and we wanted a drink and took it."[53] The group that bolted with Geronimo and Naiche numbered nineteen men, three boys, three girls, and thirteen women, including Lozen, Victorio's younger sister, who not only was as good a shot as any of the men but also had raided an army pack train and led only those mules that were carrying ammunition back to the Apache camp. According to Daklugie, Lozen had clairvoyant powers and "could locate the enemy and even tell how far away it was."[54]

Crook reported Geronimo's escape to Washington, where General Sheridan, best remembered for his statement that "the only good Indian is a dead Indian," and President Cleveland were both incensed. "I hope nothing will be done with Geronimo which will prevent our treating him as a prisoner of war," the president would later write, "if we cannot hang him, which I would much prefer."[55] Crook submitted his resignation, effective April 1. He was unwilling to go back on his word and break the terms of surrender, knowing this would confirm the suspicions of the remaining fugitives who might never give themselves up.

Miles, who had been criticizing Crook's tactics behind his back during the campaign, jumped at the opportunity not only to assume Crook's job, which he had long coveted, but also to step in and claim the capture of Geronimo for himself.

Nelson Miles was a latecomer to the American Southwest. A decorated Civil War veteran, he found his postwar ambitions frustrated, a deficiency he remedied by marrying the niece of General William Tecumseh Sherman in 1868 and assuming command of the Indian Wars. He styled himself as leading the charge in the great advance toward the West. "The buffalo, like the Indian, stood in the way of civilization and in the path of progress," he explained in his account of their mutual extermination, "and the decree had gone forth that they must both give way."[56]

The U.S. Army had been reduced to twenty-five thousand soldiers in the decade following the Civil War. Career positions were scarce. Miles had advanced his own career, one campaign at a time, by fighting against the Sioux, Cheyennes, Nez Percés, Comanches, Kiowas, and Bannocks, and, as commanding general of the Department of the Columbia, commissioned the Alaskan explorations of Lieutenant Frederick Schwatka, with an unstated goal of gaining better intelligence should the United States be faced with Native uprisings in the new territory to the north. He rose to become the last commanding general of the U.S. Army: a too-powerful position that was abolished and replaced with the position of chief of staff upon his death. The deployment by General Miles of five thousand soldiers to pursue thirty-eight Apaches, half of them women and children, was an overwhelming show of force. Finding one thousand Apaches would have been easier than finding thirty-eight.

The Geronimo campaign was Miles's chance to deploy heliography on an unprecedented scale. As fast as the army had strung telegraph lines between their outposts, Apache saboteurs became adept not only at cutting the wires in times of conflict but also at concealing the breaks with rawhide splices, making the sabotage difficult to find. Beams of light could not be cut.

Heliography, or the transmission of information by means of reflected sunlight, goes back to signal stations established among Mediterranean islands in Minoan antiquity and was first brought into modern form in 1821 by the German mathematician Carl Friedrich Gauss. Charles Babbage, the pioneer of card-programmed computing, black box data recorders for steam locomotives, and a packet-switched postal network where all destinations could be reached in the same time for the same cost, soon demonstrated that the most reliable way to encode optical signals was by the occultation of a continuously targeted beam of light. Modulation of the carrier wave became the signal, and a principle of electromagnetic telecommunication that would drive the digital revolution was given form.

Under the guidance of Henry Christopher Mance, the British army developed a portable five-inch-diameter heliograph unit with a range of about fifty miles under good conditions. These instruments, with trained operators, were deployed in India in the 1860s and in Afghanistan in the 1870s, where heliographic telecommunications were critical in supporting the British army's lengthy march through remote territory against rebels in Kabul. "The sun's rays travel ninety-two million miles to reach the earth and yet preserve vital energy enough to, even after striking a mirror, travel some score of miles farther," an early promoter of Mance's system to the U.S. military claimed.[57] Six of the Mance heliographs were acquired by the U.S. chief signal officer, General Albert J. Myer, who found no takers for them until General Miles showed up—perhaps noting parallels between the British counterinsurgencies on the Indian subcontinent and the Indian Wars in the American West. Miles established a 140-mile heliographic line between Fort Keogh and Fort Custer on the Yellowstone in Montana, where he was engaged in a protracted campaign against the Sioux and Cheyennes.

The U.S. Signal Corps, incorporating the National Weather Bureau until 1891, was responsible for training both telegraph operators and weather observers. For a heliograph operator both skill sets were combined. The assignment to Arizona was prized by the candidates at the Signal Corps training school in Virginia, who were outfitted with the latest heliograph equipment, given armed escorts to their remote stations, and supplied by pack train until provisions ran out at the end of the campaign. "We killed and ate some squirrels and the small birds we had fed all summer, until finally we waved 'adios' to Mount Baldy and started on our last trek down the mountain," reported William Neifert, one of the crew assigned to station number 8, in the Santa Rita Mountains, after the news of Geronimo's surrender came in.[58]

Portable heliograph units could be patched into the main network if a relay station was within the line of sight, while the southernmost fixed station, at San Bernardino, maintained contact by couriers with the parties who were out of sight among the canyons south of the border in search of those who were still holding out. The system worked so well that, with the Apache campaign over, the Signal Corps reconstituted the Arizona heliograph network, expanding it to fifty-one nodes covering a hundred thousand square miles, in 1891. The technology saw limited use beyond Arizona and New Mexico, although a record-breaking link was demonstrated over a distance of 183 miles between Mount Ellen, Utah, and Mount Uncompahgre, Colorado, on September 17, 1894, with Signal Corps heliographs using mirrors only eight inches square. The daily signals relayed back to headquarters at Fort Bowie during the Geronimo campaign were the distant progenitors of the bursts of code that a cell phone now transmits to the nearest tower every few seconds, day and night.

General Miles, who favored horses over mules, refused to believe the Apaches possessed skills, endurance, or weapons that the best men and equipment of the U.S. Cavalry could not match. He ridiculed Crook's methods, relegated the army mules to hauling freight, and dismissed Crook's scouts. He retained First Lieutenant Marion Maus, a veteran of the Crook campaign whose experiences in the Signal Corps aligned with the telecommunications network that Miles sought to deploy.

The Apaches, Miles admitted, possessed "a singular lung power which enabled them to climb those high altitudes without accident and with very little fatigue" and exhibited a "cunning, strength and ferocity" that had "never been surpassed in the annals of either savage or civilized crime."[59] To lead the pursuit of Geronimo, Miles selected Henry Lawton, who "was physically, perhaps, as fine a specimen of a man as could be found." According to Miles, "He weighed at that time two hundred and thirty pounds, was well proportioned, straight, active, agile, full of energy, stood six feet five inches in height and was without a superfluous pound of flesh. His bone, muscle, sinew and nerve power was of the finest texture . . . [H]e had a bright handsome face, and was in the prime of life."[60] Lawton was also an alcoholic, a weakness that seemed to have escaped notice by Miles, but not by his companions in the campaign.

As second to Lawton, Miles chose Assistant Surgeon Leonard Wood, another "splendid type of American manhood . . . a young officer aged twenty-four, a native of Massachusetts, a graduate of Harvard, a fair haired, blue eyed young man." Miles, believing that Wood "had a perfect knowledge of anatomy and had utilized this knowledge of physiology in training himself and bringing every part of his physique to its highest perfection," instructed him to "make a careful study of the Indians at every opportunity and discover wherein lies their superiority, if it does exist, and whether it is hereditary, and if hereditary, whether the fiber and sinew and nerve power is of a finer quality, and whether their lungs are really of greater development and capacity to endure the exertion of climbing these mountains than those of our best men."[61]

Wood never delivered a verdict as to whether it was lungs, sinew, or nerve fiber that enabled the Apaches to stay leaps and bounds ahead. The soldiers struggled to keep up. "The singular feature of this hard work in this high and almost impassable country is the intense craving for meat, and the immense amount everyone eats," Wood noted. "Nothing takes its place."[62] On August 4, 1886, with the search for Geronimo proving fruitless, the troops came upon some Mexican cattle, and "the largest and fattest steer which could be gotten hold of was at once shot and cut up, and we all sat around small fires roasting beef in small chunks on our ramrods or on small sticks well into the night."[63] The soldiers were forced to adopt the habits of their Apache scouts, leaving all cumbersome equipment and military tactics behind. Packmaster Henry Daly was among those sent back to headquarters during the campaign. "In case I met General Miles," recalled Daly, "I was to tell him that [Lawton] did not want any more infantrymen, and told me to say to him that he might as well try to hunt Indians with a brass band."[64]

The renegades kept evaporating into the hills, and even the scouts were unable to catch up. Contact was established only when Geronimo and Naiche chose to open the dialogue, at a place of their choosing, under their own terms. When Charles Gatewood, led by the two scouts Martine and Ka-teah, and carrying no supplies except fifteen pounds of tobacco, secured a face-to-face meeting with Geronimo, the conference opened with Geronimo's expressing his concern for Gatewood's health, which had suffered in the long campaign. "Geronimo appeared through the cane-brake about twenty feet from where I was sitting, laid his Winchester rifle down, & came forward offering his hand," Gatewood wrote in his journal. The fifty-seven-year-old warrior, twenty-four years Gatewood's senior, noted "my thinness and apparent bad health & asked what was the matter with me."[65]

General Miles was negotiating from a position of weakness and Geronimo from a position of strength. Miles was more determined to save his own reputation by delivering Geronimo than Geronimo was determined to save his own life by giving up. While Geronimo roamed freely, Miles remained confined to Fort Bowie near Apache Pass, from where he could take credit if Geronimo was killed or captured but avoid blame if he escaped. He relayed instructions by heliograph and courier to Gatewood and Lawton, orchestrating the conclusion of the campaign from a safe distance while delaying his own appearance until Geronimo and his companions were back in the United States.

After Geronimo's final surrender at Skeleton Canyon on September 4, Miles, accompanied by Geronimo and Naiche, rode back to Fort Bowie, where their arrival was greeted by the Fourth Cavalry's brass band. "They were urged to follow General Miles back to Bowie, not as prisoners of war, but as parties to an agreement made on terms to suit themselves," complained an anonymous, disgruntled army officer, "a surrender on Geronimo's own terms, made in his own camp, when he was free to do as he pleased and had Miles and his officers at his mercy."[66] Guards were posted outside the fort, less to keep the captives in than to keep curious or vengeful citizenry out. Geronimo paraded around, wearing twelve-dollar boots and a new coat and hat.

Miles knew that the terms of surrender could not be upheld. To his credit, he did succeed, against direct orders from Washington, in sending the Apaches out of the reach of the Arizona civil authorities who sought to hang them as criminals and had issued warrants for their arrest. Under his instructions, the troops escorting the prisoners to Florida protected them from the vigilantes who were trying to stop the train and execute Geronimo themselves.

The Chiricahuas were now divided. The former fugitives led by Geronimo and Naiche were now on their way, along with many of the scouts who had helped capture them, into exile in Florida; the main population was still residing on the San Carlos reservation; the subpopulation, led by Chihuahua, who had escaped in 1885 and surrendered to Crook, were already confined as prisoners of war in Florida; the remaining scouts who had been employed by the government were now back among the general population on the reserve. Additionally, a small group of ten handpicked Chiricahua leaders had been sent to Washington in July 1886, ostensibly, and at Miles's suggestion, to meet the president and come to an agreement that would relocate their entire tribe to a new reservation at a safe distance from Arizona, Miles's preference being Oklahoma, which was still Indian Territory at the time. Eleven individuals, led by Mangus, the son of Mangas Coloradas, remained at large.

The ultimate betrayal of the Apaches was the result of Miles's false promises combined with "Great Father" President Cleveland's refusal to endorse Miles's plan to relocate the Chiricahuas to tolerable rangeland in Oklahoma, insisting instead that all the Chiricahuas, guilty and innocent alike, be exiled to the intolerable swamplands of Florida. It was not Miles's intention to have the Apaches "banished to sickly Florida," but in examining the Apache problem upon taking command in April 1886, he concluded that the Chiricahuas could never be subjugated as long as they could find refuge in the mountains and desert they knew as home. He proposed "to move them at least 1,200 miles east, completely disarm them, send their stock, for the winter at least, to Fort Union, N. Mex., scatter the grown children through the industrial schools of the country, and hold the balance at one or two military posts."[67]

On July 30, 1886, Sheridan had requested authority "to immediately arrest all the male Indians now on the Chiricahua Reservation, near Fort Apache, and send them as prisoners to Fort Marion, Florida; that the delegation now here be sent there also, and that they be held at that point as prisoners of war, until the final solution of the Geronimo troubles."[68] The Chiricahuas who had been sent to visit Washington were already on board a train heading back to Arizona when orders came in to detain them until they could be sent to Florida instead. Having received an audience with the president, where Chatto, a leading scout, was presented with a medal and a certificate by the secretary of war, they assumed that their agreement to return home to live in peace would be upheld. The group led by Geronimo and Naiche, believing they were going to Florida to be reunited with their families for a two-year imprisonment, voluntarily began their journey east. The remaining Chiricahuas, some 432 individuals at San Carlos, had to be deceived and forced at gunpoint to board a specially commissioned train.

On August 20, 1886. all Chiricahua warriors were told to disarm and present themselves for inspection at Fort Apache, where armed troops encircled them. Eugene Chihuahua, thirteen years old at the time, remembers the commanding officer, Lieutenant Colonel James F. Wade, "sent word that the warriors were all to come in, and to come without any weapons. He told them that they were all to be shipped to Washington to have a visit with the Great White Father and that their wives and children were to go with them." On September 3, on the eve of Geronimo's surrender, the order was given to drive the captives to the railroad siding at Holbrook, one hundred miles away. They were forced to subsist on "the cattle that were prodded along with them."[69]

"When this caravan moved off it made a procession nearly two miles long, and contained some wagons and about 1,200 Indian ponies and, as each Indian family keeps all the dogs it can get, about 3,000 dogs," according to an eyewitness report. "As the train started, the thousands of deserted dogs tried frantically to keep abreast of the moving cars, every one howling with all his might. They were so thick, that there wasn't room enough for all of them to run, and half of them would be on the ground and the other half scrambling over them. What a sight. Gradually the dogs thinned out as the train gathered speed, but a few of them kept up for about 20 miles. The Indian horses were brought to Fort Union by the cavalry and were later sold at public sale."[70]

Eight days of confinement aboard the train were followed by twenty-seven years of imprisonment: first in Florida, where the prisoners were held in surplus military bastions dating to the time of the Spaniards, later in Alabama, and then in Oklahoma, where at last there was open sky and air. In 1913, the surviving captives, led by Daklugie, were allowed to return to Arizona and establish a settlement on the Mescalero reservation, while those who chose to stay in Oklahoma were granted title to their homes and land. They were no longer prisoners of war. A majority, numbering 187, chose to return west, and on August 25, 1971, they won a judgment of $16,489,096 against the U.S. government as compensation for the 14,858,051 acres of Warm Springs reservation that had been confiscated in 1877.

A total of 498 captives had been exiled to Florida: 99 men, and 399 women and children. Three years later, 119 had died in captivity, including 30 of the 112 children who had been removed to residential schools. "They live in terror lest their children be taken from them and sent to a distant school," Crook reported to Congress after visiting the captives in January 1890.[71] "We were accustomed to dry heat, but in Florida the dampness and the mosquitoes took toll of us until it seemed that none would be left," remembered Kaywaykla, who was about thirteen when the exile to Florida began.[72] "The only consolation we got for those terrible twenty-seven years as prisoners of war," added Eugene Chihuahua, "is that the scouts, too, were prisoners. And we made it miserable for them."[73]

At the beginning of October 1886, only Mangus and his band remained at large. They numbered three men, three women, and three children, along with Mangus's own son Frank and Geronimo's nephew Daklugie, still boys yet old enough to carry weapons and fight. Circling back into

Arizona to rejoin their relatives, Mangus's group made contact with an Apache cook working for a detachment of soldiers who were out hunting for them. Learning that all their relatives had been taken prisoner, they gave themselves up. On November 3, 1886, aboard the train carrying these last captives eastward, Mangus slipped out of his handcuffs and threw himself through a window of the thirty-five-mile-per-hour moving train. "Badly stunned [but] not seriously injured," he was recaptured, only to serve, as General Miles put it in his telegram reporting the incident, as "another illustration of the desperate efforts an Indian will make to escape."[74]

Daklugie, who estimated he was sixteen at the time, remembered the moment they surrendered to the soldiers and were disarmed. "Here's this young man; he's the most dangerous one among them," he recalled the presiding officer saying, noting his lance, which was made from a reworked saber blade attached to a wooden shaft almost twelve feet long. "When he saw that good spear of mine and the over 200 thin steel arrowheads in my quiver he understood the situation," Daklugie explained. "Those arrows would not pull out, you know; you just had to push one on through if it hit you. And it was worse than an operation today. Everybody in that camp took an interest in my arrows. I didn't mind that, but they did not give them back."[75]

These were the last arrows deployed in war against the regular army of the United States. When they were transformed from weapons into souvenirs, something was lost forever, in an instant under the October Arizona sun. The day of the bow and arrow was over. The day of the data network was coming, and the Apaches were the first to see the signs.

3

AGE OF REPTILES

I n the domain of animals, cold-blooded preceded warm. In the domain of electronics, warm-blooded preceded cold.

It took less than one hundred years to make the leap from the coded pulses of light transmitted through the heliograph network deployed against Geronimo to the coded pulses of light transmitted through the fiber-optic telecommunication networks of today—as if the clear, desert air of the Arizona territory had somehow been preserved, encapsulated, and insinuated to the far reaches of the globe. Vacuum tubes were a step along the way.

The dawn of modern electronics goes back to John Ambrose Fleming's puzzlement over the "Edison effect": a suite of odd behaviors first evidenced in 1880 by Thomas Edison's investigation of the secondary conductivity within an evacuated incandescent lamp. In driving an electric current through a carbonized filament to produce light, Edison had unintentionally liberated electrons into the vacuum, an escape that might have gone unnoticed except that along with these electrons went traces of material eroded from the filament and deposited unevenly on the inner surface of the glass.

This erosion of the filament shortened the life of bulbs and revealed a puzzling asymmetry: the incandescent filament radiated *light* in all directions, while the eroded *particles* favored one side of the lamp. Edison introduced a third electrode into a series of experimental lightbulbs to probe what was going on inside the vacuum and, connecting a galvanometer, found that when the primary filament was energized, the vacuum would conduct electricity, but only in one direction. The vacuum was behaving as a *semiconductor*, and the stage was set for a revolution that would outshadow electric light.

Edison failed to grasp the implications of his discovery, seeing only the immediate utility of producing an "indicating" lightbulb that could help regulate an electrical distribution network by signaling, on a secondary circuit, the state of the current at the location of the device. "I have discovered that if a conducting substance is interposed anywhere in the vacuous space within the globe of an incandescent electric lamp," he wrote in a patent application filed on November 15, 1883, "a portion of the current will, when the lamp is in operation, pass through . . . the vacuous space within the lamp."[1]

Fleming joined the Edison Electric Light Company of London as scientific adviser in 1882. He was alerted to Edison's discovery in October 1884 by William Henry Preece, chief engineer of the British Post Office, who brought news of "a very striking experiment with glow-lamps," after a visit to Edison's laboratory, "the principle of which [Edison] had not threshed out."[2] Fleming elaborated upon Edison's experiments, asking questions where Edison had left off. "This new and interesting effect became known as the 'Edison effect,'" he reported, "but Mr. Edison gave no explanation of it and made no practical application of it in telegraphy, or for any other important purpose."[3]

A peculiarity of Edison's indicating lamp was that if the primary filament was energized by an alternating current, this produced a non-alternating, *rectified* secondary flow, as if the lamp were treating the electric current like a crowd of people at a subway turnstile, only allowed to flow one way. The phenomenon, reported by Fleming in 1896, was dismissed by Edison as having no practical application to the lighting and power distribution systems that dominated his commercial interests at that time.

Fleming was born in 1849, the son of a Lancaster minister and grandson of the Kent-based portland cement manufacturer John Bazley White, "a man of original mind who, early in the nineteenth century, carried out his own conception of building his residence entirely of cement."[4] Fleming worked as a science teacher, draftsman, and research chemist before entering St. John's College, Cambridge, under a scholarship in 1877. Drawn to physics, he joined the Cavendish Laboratory, at times the only student in attendance at James Clerk Maxwell's lectures until Maxwell's death in 1879. It was Maxwell who encouraged Fleming to investigate the induction of electric currents in gases, leading to Fleming's interest in conductivity within the vacuum as a special case. Fleming, who popularized the term "electronic," helped initiate the practical reconciliation of the grand, top-down unification of Maxwell's equations, governing the general behavior of electromagnetic fields, with the peculiar behavior of individual electrons that was coming into evidence from the bottom up.

Fleming was also a passionate anti-Darwinist, believing the "reckless popularization of the theory of organic evolution, without regard to the strong arguments which can be urged against it," to be "disastrous to the ethical development or spiritual life of the young."[5] These arguments were collected in his 1935 treatise *The Origin of Mankind: Viewed from the Standpoint of Revelation and Research*. According to Fleming, both scripture and science, from Genesis to quantum mechanics, revealed a common truth. "We have to regard the Universe," he concluded, "not as a collection of Things or Events existing apart from any awareness of them by observers, but as manifested Thoughts in a Universal Mind."[6]

During his tenure with Edison Electric, Fleming promoted the adoption of electric lighting by industry, the general public, and the Admiralty. He wrote the first textbook on power transformers and was called in to investigate a series of accidents that plagued the London electrical distribution system, tracing the cause of arcing across the low-voltage switchboards to currents induced by stray electromagnetic fields from the high-voltage side. He maintained his teaching position and research laboratory at University College London, and with access to the lamp-making facilities of Edison and later Edison Swan Electric, he was able to continue his study of the Edison effect.

On February 14, 1890, Fleming delivered a lecture to the Royal Institution of Great Britain titled "Problems in the Physics of an Electric Lamp." When the filament was energized, "the surface molecules are shot off in straight lines, and . . . are charged, if at all, with *negative electricity*," he explained, demonstrating asymmetric conductivity through the vacuum and noting that "there must be some way by which negative electricity gets across the vacuous space from the negative leg of the carbon to the metal plate, whilst at the same time a negative charge cannot pass from the metal plate across to the positive leg." As "the beginnings of a theory proposed to reconcile these facts," he suggested "that the interior of the bulb of a glow-lamp when in action is populated by flying crowds of carbon atoms all carrying a negative charge."[7]

In March 1896, still puzzling over the variable conductivity of the vacuum, Fleming observed that "sending a small current through this space seems to effect a change in the qualities of the rarefied gas . . . which makes it conduct better."[8] One year later, Joseph John Thomson identified the subatomic particles that were carrying the electric charge, which George Johnstone Stoney, second cousin to one of Alan Turing's great-grandfathers, had already termed "electrons" in 1891. Within the vacuum these unbound electrons obeyed their own laws, unobservable to us, until they either reentered the world of metal, producing a detectable current, or shifted state through collision with other atoms, producing photons visible as light.

In 1899, Fleming signed on, at three hundred pounds per year, as scientific adviser to Guglielmo Marconi, the wireless pioneer. Marconi had arrived in London at the age of twenty-two in 1896, bankrolled by a syndicate of distillers led by the Jameson family of Ireland in a gesture of reconciliation with his mother, the heiress Annie Jameson, who had eloped, in defiance of her father, to marry Giuseppe Marconi, a Bolognese silk merchant seventeen years her senior, at age twenty-five in 1864. Annie Jameson, who accompanied her son to London to secure the investment and help manage his affairs, had nurtured his genius from an early age. "If only grownups understood what harm they can do to children," she later advised. "They think nothing of constantly interrupting their train of thought."[9]

Marconi was a gifted technician, inventor, and master of the product demonstration, but it was Fleming's understanding of Maxwellian electrodynamics and experience with high-voltage generation and distribution systems that boosted Marconi's radio transmissions to transatlantic strength. Marconi also needed the endorsement of someone within the British academic establishment, and Fleming, a popular lecturer associated with the legacy of James Clerk Maxwell, fit the bill.

In December 1900, Major Flood-Page, managing director of Marconi Wireless, responded to Fleming's proposal for a transatlantic experiment, which Marconi referred to as the "Big Thing," with a raise to five hundred pounds per year, while reminding him that "the main credit will be and must forever be Mr Marconi's."[10] Marconi personally added a promise of five hundred shares in Marconi Wireless "in the event of our being able to signal across the Atlantic," adding that the shares "will be very valuable if we get across."[11]

Fleming supervised the design and construction of Marconi's long-range transmitter at Poldhu on the outer Cornwall coast. His quest to deliver the maximum-possible electromagnetic disturbance left no principle of physics or engineering unturned. A thirty-two-horsepower Hornsby-Akroyd hot-bulb oil engine drove a twenty-five-kilowatt Mather & Platt alternator whose output was raised by a series of transformers to twenty thousand volts. A second-stage accumulator boosted the output to a hundred thousand volts, sufficient to generate a two-inch spark. All control switching was immersed in oil to suppress arc. After Fleming returned to London in September, Marconi himself tuned the transmitter to drive a two-hundred-element antenna, suspended in the form of an inverted cone between a circular array of two-hundred-foot-tall wooden masts. The structure, with running topmasts and a network of running stays, was rigged as if five sailing ships with four masts each had formed a circle and run aground.

On September 17 the antenna collapsed in a storm and was reduced to a fifty-four-element array supported by two surviving spars. After the transmitter was retuned, Marconi departed for Canada, leaving instructions for the Poldhu station to transmit the letter _S_ in Morse code repeatedly at scheduled times. Marconi, assisted by George S. Kemp and Percy Wright Paget, hoisted a six-hundred-foot aerial by kite on Signal Hill near St. John's, Newfoundland, and claimed to have detected the signals, at a distance of eighteen hundred miles, on December 12 and 13, 1901.

In Morse code, _S_ consists only of three dots. Was the _S_ received in Newfoundland really the _S_ sent from Cornwall, or an artifact picked up out of the ether at the prearranged time? Marconi was using a sensitive but temperamental detector consisting of a globule of liquid mercury sealed inside a glass tube with iron contacts at each end. These electrodes had to be touching the surface of the mercury, but not quite making full wetted contact, to detect a signal, acting as a rectifier when the partial contact was just right.

There was no independent confirmation of the transmission, and Marconi's claims were met with skepticism—and a lawsuit from the Anglo-American Telegraph Company, which held an exclusive license to convey transatlantic communications through Newfoundland. Its shares dropped precipitously in response to the news from St. John's. "One swallow does not make a summer, and a series of 'S' signals do not make the Morse code," *The Daily Telegraph* complained.[12] The Canadian government then weighed in, inviting Marconi to establish a radiotelegraph facility in Nova Scotia instead of still-British Newfoundland. Marconi sidestepped the lawsuit and secured the investment he needed to stay afloat.

After a year of estrangement over the apportionment of credit for the transatlantic experiment, Fleming returned to Marconi's enterprise. He shifted his attention from the power of the transmitter to the sensitivity of the detector, where he was able to make use of the Edison effect. The difficulty in detecting radio waves, even at long wavelengths and low frequencies, is that the current they induce in an antenna oscillates thousands of times per second. Averaged over any perceptible period of time, the voltage fluctuations cancel out. By filtering these oscillations through a matched pair of Edison indicating lamps, Fleming produced a continuous current that was far easier to detect.

"These bulbs have the property that inside the bulb, negative electricity can move from the hot to the cold carbon filament even under very low electromotive force but it cannot move in the opposite direction," Fleming reported in a patent application filed on November 16, 1904. "All the positive alternations pass through one bulb and all the negative through the other bulb."[13] The resulting signal could be rendered visible by a mirror galvanometer such as had already been developed for revealing the weak impulses received over submarine cables of transatlantic length. Fleming wrote to Marconi disclosing this "interesting discovery . . . a method of rectifying electrical oscillations . . . so that I can detect them," but he kept the patent application to himself. "I have not mentioned this to anyone yet as it may become very useful" was all he said.[14]

Fleming named the new device an oscillation "valve," "because it acts towards electric currents as a valve in a water-pipe."[15] Visual display of a detected signal was preferable to audible, he added, "because if a code message was being received, the omission of a single dot or dash might make nonsense of a word." It was an accident of history that Fleming himself was going deaf and was compelled "to find some instrument to record radiotelegraphic signals which would appeal to the eye and not the ear."[16]

After a series of demonstrations using off-the-shelf Edison indicator lamps, Fleming commissioned a proprietary version, marketed as the "Fleming valve." This was no less of an invention for being the application of a device that already existed for one purpose to something else, but Fleming's patent claims were judged to have been injured by prior art.

Fleming's thermionic valve, or vacuum tube, as it came to be known in America, operated as a doorway for electrons, without *doing* anything to those electrons other than letting them through if they were going in the right direction, and shutting them out if not. It was an American contemporary of Fleming's, Lee de Forest, who introduced a third, controlling electrode into Fleming's valve. De Forest bent a piece of platinum wire into the shape of a fireplace gridiron and sealed it inside a Fleming valve to form a *triode*, with the third electrode known ever after as the grid. This enabled not only the detection but also the *amplification* of a signal, launching the age of electronic communication and control.

De Forest refused to acknowledge the extent to which his own invention was based on Fleming's, while Fleming, who accused de Forest of misappropriation, admitted his own oversight in not adding a control grid to his valves. The two inventors were opposite in their approach. Fleming, the protégé of Maxwell, developed his inventions on a sound theoretical foundation in well-equipped laboratories and allowed the Edison and Marconi organizations to take much of the credit for their success. De Forest, whose overtures to Nikola Tesla were rebuffed, experimented independently without any consistent explanation as to how his inventions were supposed to work, and when they did work, he outmaneuvered his investors to preserve his own undiluted interests and was even accused of fraud. Instead of attaching himself to an Edison or a Marconi, de Forest conducted much of his early research in his bedroom in a low-rent Chicago rooming house while supporting himself by working in a factory that made dynamos and periodically "walking about Chicago trying to find an edible fifteen-cent steak."[17]

Until the invention of the vacuum tube, detecting radio signals at long distances remained a black art. The radiotelegraph operator relied either on a crystal detector invoked by tickling the surface of a small fragment of naturally occurring semiconducting mineral, such as lead sulfide or iron pyrite, with a very fine metallic whisker until a point of contact constituting a rectifier was found, or on a "coherer" consisting of a glass tube filled with loosely packed iron filings that coalesced under a radio frequency impulse and then had to be tapped back into a nonconducting state before the next bit of incoming information could be read. The mechanical sensitivity of these detectors was also their downfall, because stray vibrations could bring reception to a halt. What was needed was a detector sensitive to radio frequency oscillations but resistant to mechanical shock.

Puzzled by a gaslight in his rooms that flickered during radio transmission tests, de Forest speculated that enveloping two electrodes, separated by a small gap, within an active flame could "enable them to act as a detector of electrical oscillations [because] such condition or molecular activity causes what would otherwise be a non-sensitive device to become sensitive to the reception of electrical influences." He applied for, and eventually secured, a patent for "a self-restoring, constantly-receptive oscillation responsive device comprising in its construction a sensitive conducting gaseous medium," which was never shown to work, and he later admitted that it was sound waves, not radio waves, that caused the flickering of the lamp.[18] "The illusion had served its purpose," he explained. "I had become convinced that in gases enveloping an incandescent electrode resided latent forces, or unrealized phenomena, which could be utilized in a detector of Hertzien oscillations far more delicate and sensitive than any known form of detecting device."[19]

In late 1901, de Forest established the American de Forest Wireless Telegraph Company, signing up investors to develop a series of modified

Fleming valves. He christened his invention the "Audion," but it failed to live up to its name as a detector of radio signals, and the growing wireless industry passed it by. On November 25, 1906, he delivered to H. W. Mc-Candless, his tube fabricator, the drawings for a new design, "comprising an evacuated vessel, two electrodes, one of which is a filament, inclosed within said vessel, means for heating said filament, and a grid of conducting material inclosed within said vessel and interposed between said electrodes."[20]

Three days later, de Forest submitted his resignation from the company he had founded, taking with him a thousand dollars and the apparently worthless Audion patents as severance pay. According to de Forest, not until December 31 did his assistant, sixteen-year-old John Vincent Lawless Hogan Jr., get around to retrieving the prototype "grid-type" Audions from McCandless and trying them out. They worked. De Forest had only a vague idea *why* they worked, but this was no longer an obstacle to success. "The explanation of this phenomenon is exceedingly complex and at best would be merely tentative," he wrote in his disclosure of the invention, and "[I] do not deem it necessary herein to enter into a detailed statement of what I believe to be the probable explanation."[21]

Some inventions result from theorizing how something should work and then building it. Others result from building something that works before understanding it. When de Forest's Audion entered the market in 1907, there was still no coherent theory of what was going on inside the vacuum. Electrons had been harnessed, controlled, and collectively put to work within wires, batteries, electromagnets, and incandescent filaments without paying any attention to their existence as individuals. Free electrons were only acknowledged collectively, as "cathode rays."

Within the vacuum tube, the flow of electrons from cathode to anode could be controlled, without intermediate mechanical steps, by a much smaller current applied to the grid. An otherwise undetectable signal, originating hundreds or thousands of miles away, could be amplified into a detectable and ultimately audible flow. The world huddled around its radio sets, and the electronics industry was born. By 1940, more than a hundred million vacuum tubes a year were being produced in the United States. You could buy them at the local drugstore where self-service tube-testing machines allowed consumers to determine whether a suspect tube really needed replacement or not.

Vacuum tubes, with no moving parts except electrons, enabled machines to operate at the speed of light and frequency of radio waves, unlike mechanical devices limited by the speed and frequency of sound. At first the flow of electrons was controlled the way a valve controls a flow of fluid through a pipe, but the degree of control soon expanded: in time, by building tubes that produced pulses of electrons rather than a continuous stream; and in space, by building tubes that utilized the deflection of an electron beam by an electromagnetic field. Cathode-ray tubes and image tubes were introduced, able to translate in both directions between the internal world of captive electrons and the visible and invisible spectrum of electromagnetic radiation ranging free in the world outside.

Along with audio equipment delivering higher fidelity than the digitally sampled approximations that dominate today, vacuum tube development culminated in three mid-twentieth-century technologies: television, which dominated the peace; radar, which dominated the war; and electronic digital computers, which dominated both. The television camera tube scanned a moving image in real time, capturing it as an analog waveform broadcast over the air and reconstituted by distant cathode-ray-tube displays. Radar (radio detection and ranging) systems scanned the horizon with a pulsed beam of radio waves, thousands of watts in strength, yet were able to identify a moving target by distinguishing an echo amounting to as little as a trillionth of a watt. Electronic digital computers enabled bits of information whose transformation had been governed by the speed of electromechanical switching to escape these limitations and shift both logical state and physical location from one microsecond to the next.

Until the age of silicon, electronics was dominated by vacuum tubes and cathode-ray-tube displays. Electronics was tangible in ways that are lost to us today. You could see the heater elements glowing through the ventilated cabinets of radio and television sets, and in certain tubes that were not pure vacuum, there was a ghostly flicker as the trace gas inside the tubes fluoresced. Power supply and fly-back transformers hummed while the faces of cathode-ray tubes, bombarded by electrons, accumulated sufficient charge to deflect a silk blouse worn by someone walking by. Vacuum tube equipment emitted a characteristic odor that was partly the signature of ozone produced by static discharge and partly the scent of ambient dust particles that were attracted, like moths to a flame, by surfaces too hot to touch.

Nowhere in postwar America was there more electronics, and better electronics, than in New Jersey. From RCA's vast Camden works and Princeton laboratories to the Bell Telephone Laboratories in Murray Hill, it was New Jersey that led the way in electronics just as Thomas Edison, and the city named after him on the shores of the Raritan River, had led the way in delivering electric light.

Princeton, New Jersey, was only twenty miles from Edison, yet a world apart. Founded by a syndicate of Quaker families fleeing the eighteenth-century gentrification of Philadelphia and New York, it remained protective of its prerevolutionary architecture, surrounded by farmland reverting back to forest or undergoing development now that food could be grown more efficiently at refrigerated distances from New York. Its central feature was a university, founded as the College of New Jersey in 1746, with a number of laboratories and institutions in its orbit, some devoted to the cultivation of ideas and others devoted to the development of stuff.

The laboratory most devoted to stuff belonged to the Radio Corporation of America (RCA). Successor to the American Marconi Company and parent to the National Broadcasting Company (NBC), RCA was established under a government order to keep the radio industry, nationalized during the wartime emergency, under U.S. control in the aftermath of World War I. RCA, and its subsidiary RCA Victor, formed through a merger with the Victor Ta king Machine Company, whose trademark fox terrier, Nipper, was adopted by RCA in one of the most successful branding exercises of all time, became the dominant provider of both radio equipment and programming: a vertical monopoly that survived the transition from radio to television without missing a step. RCA, masterminded by the Russian immigrant David Sarnoff, was also the progenitor of a trend among American technology companies to separate their research facilities, requiring a permanent staff of well-compensated scientists, from their manufacturing facilities, located wherever labor was available at the lowest cost.

RCA's Princeton laboratories, forty miles from the dense, low-income Camden neighborhoods that were home to the ten thousand workers who built the products that the engineers in Princeton designed, were established in 1941, shortly after a period of violent labor unrest at the Camden plant. The 2.5-million-square-foot Camden facility was supplied with raw materials, ranging from coal for its generating station to lumber for radio cabinets, by its own railroad and barge terminals. Up to five thousand radio sets per day were shipped out.

The institution most devoted to ideas was the Institute for Advanced Study (IAS). Proposed in 1924 by the Norwegian American topologist Oswald Veblen, whose uncle Thorstein had coined the term "conspicuous consumption" in his 899 *Theory of the Leisure Class*, the institute was founded in 1930 through the generosity of the Newark dry-goods merchant Louis Bamberger and his sister Carrie (Mrs. Felix Fuld) and launched into the aftermath of the Great Depression by the high school teacher turned educational reformer Abraham Flexner, who described the thinking behind his project in an essay for *Harper's Magazine* titled "The Usefulness of Useless Knowledge" in 1939. The Bambergers had retired to devote themselves to philanthropy, selling their operations to R. H. Macy & Co. just weeks in advance of the 1929 stock market crash. The result was an educational institution that neither required advanced degrees nor awarded them. It was intended, as Flexner described it, to serve as "a paradise for scholars who, like poets and musicians, have won the right to do as they please."[22]

The scholars were divided into two classes: visiting members in residence, with rare exceptions, for one year or less, and permanent members, in residence for life. Salaries were kept high for the permanent members so they would stay, while stipends were kept low for the visitors so they would not. No reports were required, and no classes were taught. Committees and faculty meetings were outlawed because, according to Flexner, "once started, this tendency toward organization and formal consultation could never be stopped."[23]

The institute, whose first appointments were Flexner, Veblen, and Albert Einstein, opened with a school of mathematics, followed by a school of history, including archaeology and art, and a short-lived school of economics and politics that the Bambergers hoped would "contribute not only to a knowledge of these subjects but ultimately to the cause of social justice which we have deeply at heart."[24] Physics and astronomy remained under the wing of mathematics until the School of Natural Science was established in 1966. During World War II most of the physicists, with the exception of Einstein, left to join the Manhattan Project at Los Alamos or its ancillary labs. Everything was different when they returned to academic life.

The atomic bomb was analogous to Lee de Forest's Audion in that it was made to work in advance of a complete understanding of what had happened at the energy densities unleashed by the explosion of the device. With the war over, theoreticians who had dropped everything to build the weapon picked up where they had left off. Robert Oppenheimer returned to Berkeley, Enrico Fermi returned to Chicago, and Hans Bethe, bringing Richard Feynman with him, returned to Cornell. Skills, equipment, and facilities that had been developed during the war were now available for fundamental research, and the theoreticians were flooded with new, unexplained, and sometimes contradictory experimental results.

One of these results, known as the Lamb shift, was an almost imperceptible anomaly in the spectrum of atomic hydrogen: a puzzle, like the Edison effect, that was observed long before it was explained. The hydrogen atom, consisting of one proton and one electron, was both the most common and, mathematically, the best described physical object in the universe. The spectacular success of the new theory of quantum electrodynamics rested on a precise description of the hydrogen atom, but when examined with the new techniques that had emerged in radar laboratories during the war, the hydrogen atom did not behave exactly as Paul Dirac's theory prescribed. The fine structure of its energy spectrum was shifted slightly from where it should be, and this discrepancy was named, in honor of its measurement, after Willis Lamb.

During World War II, Lamb, being married to a German national, was disqualified from joining his colleagues at Los Alamos but, after months of delay, received clearance to work at Columbia University's radiation laboratory, a satellite of the radar development work centered on Cambridge, Massachusetts, under the auspices of MIT. He was assigned to work on K-band microwaves, which, at a wavelength of 1.25 centimeters, suffered interference from water vapor in the atmosphere, marked by a prominent absorption line in the radiation spectrum: a frequency at which the target substance absorbs energy and changes state. For water and K-band microwaves, this results in the water heating up. This would lead both to the microwave oven, which takes advantage of this interference, and to Lamb, in his investigation of the problem, helping to develop the new field of microwave spectroscopy from scratch.

Before the war, using optical spectroscopy, researchers had observed certain absorption lines of the hydrogen spectrum to be smudged or shifted slightly, but it was impossible to determine if they really were shifted, and, if so, by how much. After the war, Lamb, working with Robert C. Retherford at Columbia University, obtained a precise measurement of the difference in energy between two metastable states of hydrogen that according to Dirac's theory should have been identical but were not. The experiment was a work of art both in its theoretical concept and for the technical ingenuity required to identify, separate, and detect the one out of a hundred million atoms that, for a small fraction of a second, could be coaxed into the desired metastable state.

Results began coming in on April 26, 1947, and, as Lamb remembers, "it was obvious from the first indication of resonant transition that discrepancies from the predictions of the Dirac theory of hydrogen were being observed and that they were far larger than then thought."[25] Lamb was awarded the Nobel Prize in 1955. For both the name of the phenomenon *and* the prize to go to the person who *measured* the shift, independent of the prizes later awarded for the theory that explained it, is a testament to the importance of the numerical discrepancy revealed by Lamb.

Sometimes, theorists come up with a number, and the experimentalists are left to ask Nature whether the number is correct. At other times, experimentalists come up with a number, and the theorists are left to explain the number that Nature has already shown to be correct. For the Lamb shift, Nature came first, and the theorists were left scrambling to catch up. "Everybody was highly euphoric," remembers Julian Schwinger, who attended the American Physical Society meeting at Shelter Island, a resort near the end of Long Island in New York, where the Lamb results were officially announced. "Everybody was talking physics [again] after five years. The facts were incredible: to be told that the sacred Dirac theory was breaking down all over the place!"[26]

Hans Bethe began a rough nonrelativistic calculation on the train ride back to Schenectady from Long Island. At least four other teams were soon working to extend his calculation, including Lamb himself (with Norman Kroll), Victor Weisskopf (who had already been working on the problem, with Bruce French, before Lamb's announcement), Enrico Fermi, Julian Schwinger, Richard Feynman, and Bethe's graduate student Richard Scalettar. Reconciling the difference between the theoretical predictions and the experimental observations was the most pressing problem in physics, and it was worth the risk of duplicated effort to pursue multiple approaches at once.

In September 1947, these efforts were converging upon a range of improving but still conflicting approximations when a twenty-three-year-old British graduate student named Freeman John Dyson showed up. He arrived at Cornell on a postwar Commonwealth Fellowship with an introduction from G. I. Taylor, of Cambridge's Cavendish Laboratory, who had worked with Bethe at Los Alamos during the war. Dyson was assigned yet another variant of the Lamb shift calculation: ignoring the electron's spin but taking relativity into account.

"[Bethe] has given me a definite research problem to work on which is very much of the sort that I need," Dyson reported to his parents six days after arriving at Cornell, "a lengthy calculation . . . which has considerable theoretical interest although it concerns an unobservable phenomenon."[27] The difficulty with the Lamb shift calculations was that all the attempts to quantify the self-energy of the electron diverged into infinities that led to physically meaningless results. Bethe's initial calculation disregarded these infinities, arriving at a first approximation of the value observed by Lamb, but the consensus among physicists was that the troublesome infinities had been swept under the rug. Dyson, fascinated by infinities since childhood, wrangled them head-on.

"I don't know how old I was; I only know that I was young enough to be put down for an afternoon nap in my crib," he says, remembering how he discovered a convergent infinite series for the first time. "The crib had solid mahogany sidepieces so that I couldn't climb out. I didn't feel like sleeping, so I spent the time calculating. I added one plus a half plus a quarter plus an eighth plus a sixteenth and so on, and I discovered that if you go on adding like this forever you end up with two. Then I tried adding one plus a third plus a ninth and so on, and discovered that if you go on adding like this forever you end up with one and a half. Then I tried one plus a quarter and so on, and ended up with one and a third. So I had discovered infinite series. I don't remember talking about this to anybody at the time. It was just a game."[28]

In September 1932, at the age of eight, he was sent to Twyford School. "It was an abominable school but had an excellent library so that was my refuge," he remembers. He began to teach himself physics. "There was lots and lots of stuff about electrons and electricity and radio waves and all sorts of things, but nobody ever mentioned protons and I couldn't figure out why," he says. "I remember asking people, 'Why is it that they only talk about electrons and not about protons?' Nobody seemed to know." At Twyford, he organized a raffle, taking in small amounts of money from other students, paying half of it back to the winner, and pocketing the rest. "Yes, they knew I was keeping half of it," he admits, "but they kept giving me their money!"

He won a scholarship to Winchester College at age twelve in 1936, then to Trinity College at Cambridge in 1941. He owed both his name and his existence to his maternal uncle Freeman Atkey, his father's best friend, teaching colleague, and fellow motorcycle enthusiast who had been killed by a sniper in World War I. Dyson's parents, George Dyson and Mildred Atkey, were brought together by their shared loss. World War II, in contrast, suited the seventeen-year-old who arrived at Cambridge in October 1941 just fine. The older students and younger professors had left, leaving the remaining students alone with senior professors who included legendary figures in mathematics and physics such as G. H. Hardy, J. E. Littlewood, Arthur Eddington, and Paul Dirac. During Dyson's first term at Trinity, Hardy was lecturing on Fourier series to an audience of four. "There is no talk of examinations or anything silly like that."[29]

During his accelerated course of study at Trinity College, Dyson slept, while on fire patrol duty for the Home Guard, in a bomb shelter at night. On March 23, 1943, he set off by bicycle from Cambridge to the nearby town of Ely, passing a wartime aerodrome along the way. "The bombers, which I believe were Halifaxes though I am not sure," he reported to his parents, "took off directly over the road; one of them came over about 20 feet above my head . . . They were all painted jet black underneath, so they were evidently intended for use at night."[30]

In July 1943 Dyson's formal education came to an end. After the loss of an unacceptable number of British scientists in World War I, it had been decided to avoid sending scientists into combat in World War II. He declined the offer of "a job in the d-c-d-ng department (this I am not supposed to divulge)" while adding that an assignment to assist in the code breaking at Bletchley Park would be "not so bad as a computing post."[31] After an interview with C. P. Snow, he was assigned to the Royal Air Force Bomber Command's Operational Research Section in Buckinghamshire, where his job as a statistician was to improve the effectiveness of the attacks against Germany for which the bombers he had seen four months earlier had been painted black. He lived and worked among the flying crews, mostly boys his own age or younger, who faced a one-in-twenty chance with every mission of not coming back.

The reasons for the loss of so many British aircraft were poorly understood. Flying closer together increased losses to collision with friendly aircraft, while flying farther apart increased losses to enemy attack. Pilots preferred to die by enemy fire than through collision, and their behavior was difficult to change. The RAF bosses also wanted to know why so many German factories survived repeated bombings and why the firestorm that destroyed much of Hamburg in 1943 proved so difficult to reproduce. Dyson would later be sued for libel when he published an insider's account of the campaign to inflict maximum loss of life.

When the war ended, Dyson took a job at Imperial College London, followed by a fellowship to Trinity College at Cambridge, where he shifted his attention to mathematical physics and decided to visit the United States. He arrived in New York harbor aboard the Cunard liner *Queen Elizabeth* on September 17, 1947, at 6:30 a.m. Bethe, who had led the Theoretical Division at Los Alamos, had transplanted its wartime camaraderie to the physics department at Cornell. The Lamb shift had left physics in disarray. "If you don't understand the hydrogen atom, you don't understand anything," Dyson remembers, "and to find that things were wrong even with a hydrogen atom was a big shock."[32]

In 1947 physics was still in transition from the old physics in which things gave rise to fields to a new physics in which fields gave rise to things. America had taken the lead in the experiment-driven study of elementary particles, but Europe remained ahead in the mathematics-driven understanding of fields. At Cambridge in 1946, Dyson's adviser, Nicholas Kemmer, had given him a copy of *Einführung in die Quantentheorie der Wellenfelder* (*Quantum Theory of Fields*) by Gregor Wentzel, published in Vienna in 1943. "It was at that time a treasure without price," says Dyson. "I believe there were then only two copies in England."[33] He showed up in Ithaca, New York, with a working knowledge of quantum field theory at just the right time.

"Bethe and Feynman," remembers Dyson, "had been doing physics successfully for many years without the help of quantum field theory."[34] They didn't pay much attention as he set to work. A few weeks later he produced an answer that was in close accord to the numbers observed by Lamb. The attitude changed. "The convergence of the present calculation is noteworthy and somewhat unexpected," he reported in a paper that Bethe shepherded into print.[35] "Bethe was impressed," Dyson adds. "He said it was the first time he had seen quantum field theory do anything useful."[36] The simplified model, while disregarding spin, was "at least conceptually a physical system (which a non-relativistic system is not) [and] gives a convergent expression . . . very close to the non-relativistic approximations and to the observed shifts."[37]

Dyson became Feynman's mathematical ally and personal friend. Feynman, who had grown up in Far Rockaway in Queens, instructed Dyson not only in quantum electrodynamics but also in the arcana of American life. They took long road trips together, and on a drive to Albuquerque in 1948, Dyson fell in love with the American West. Besides "his own private version of quantum theory," as Dyson described it, Feynman had an almost supernatural knack with women, which Dyson, having grown up in the all-male British educational system, lacked.

In January 1949 they were attending the annual meeting of the American Physical Society in New York when a speaker mentioned "the beautiful theory of Feynman-Dyson" in reverent tones. "Feynman turned to me," Dyson reported to his parents after the meeting, "and remarked in a loud voice, 'Well, Doc, you're in.'"[38]

Being "in" included being invited to join Robert Oppenheimer at the Institute for Advanced Study. In the aftermath of the Manhattan Project, the IAS, although publicly associated with nuclear physics because of Einstein, had been left behind. The institution that had done so much to bring nuclear physics to America through its behind-the-scenes role, in collaboration with the Rockefeller Foundation and the Emergency Committee in Aid of Displaced German Scholars, in spiriting key nuclear physicists out of danger in Europe before World War II, was now outshadowed by the national weapons labs that were turning their attention to fundamental research.

In January 1946 the IAS had proposed using a portion of its own eight-hundred-acre site to build a new, East Coast nuclear physics laboratory to serve as a less remote version of Los Alamos. This plan was met with opposition from the faculty in general and from Einstein in particular, who feared that experimental physics might draw the IAS into weapons-related work and "further ideas of 'preventive' wars."[39]

Lewis L. Strauss, chairman of the IAS board of trustees and later chairman of the U.S. Atomic Energy Commission (AEC), saw a path forward and pushed it through: offering Oppenheimer, the wartime director of Los Alamos, the directorship of the institute and encouraging him to recruit a small circle of brilliant young theoretical physicists such as he had assembled at Berkeley before the war. Oppenheimer took the lure.

Robert, his wife, Kitty, and their two children moved into Olden Manor, the director's residence that overlooked the grounds of the former Olden Farm. A connoisseur of art, literature, and history as well as science, Oppenheimer reinvented the institute as an "intellectual hotel."

In addition to the scholars already in residence, he added luminaries like the child psychologist Jean Piaget and the poet T. S. Eliot, who was invited for the fall term of 1948. The entire group lived more or less communally: individual scholars in nearby rooming houses, visiting members with families in a complex of war-surplus housing that had been brought in by rail from a decommissioned iron mine in upstate New York, and the permanent faculty in nearby homes that were being built on vacant institute land. Meals were served in a cafeteria on the top floor of Fuld Hall, the headquarters that had been constructed in 1939. Afternoon tea, a custom introduced by Oswald and Elizabeth Veblen, was served in the common room on the ground floor at three o'clock. "Tea," in Oppenheimer's description, "is where we explain to each other what we do not understand."

Einstein and Oppenheimer occupied offices in opposite wings of Fuld Hall. Einstein convened the Emergency Committee of Atomic Scientists, calling for international control of all nuclear weapons, while armed guards stood watch over Oppenheimer's U.S. Atomic Energy Commission safe. It was an open secret that the IAS Electronic Computer Project, launched in late 1945 and housed in the Fuld Hall basement before moving to a nondescript outbuilding, was being rushed into existence to perform the numerical hydrodynamics necessary to design hydrogen bombs. There would be a thousandfold increase in both explosive yield and computing power between the fission weapons dropped on Japan in 1945 and the thermonuclear Castle Bravo test of 1954.

Postwar euphoria was in the air. Foodstuffs, building materials, fuel, and housing were still rationed, but these hardships were coming to an end. Those who had survived the war now made up for lost time. Oppenheimer, as Dyson reported to his mother after the first dinner he attended at Olden Manor, "looked happier than I have seen him ever" when T. S. Eliot, who would list *The Cocktail Party* as his only publication related to his institute residence, was presented to the guests. "When [Oppenheimer] spoke to me," Dyson continued, "it was to give me the recipe for some delicious Mexican savouries that were being served with the supper (he is an expert cook); then he rushed off to the next conversation, which might have been on any subject from football to cuneiform texts."[40]

In the spring of 1949, *The New Yorker* sent its Talk of the Town correspondent on the train down to Princeton to bring back a firsthand report. The hydrogen bomb debate had not yet emerged into public view, and the picture was of an untarnished paradise, only an hour from New York.

"We saw Oppenheimer from a distance, apparently engaged in being wheedled out of a nickel by a small child," *The New Yorker* reported. "Under our friend's guidance, we peeked in on a dozen offices in one of the Institute's three red brick college buildings, and were introduced to several thinkers occupied with their labors. One was doodling on a scratch pad, presumably developing a momentous equation; one was staring out the window at four crows in a field; and a third was writing a letter to his mother on a portable typewriter." In between a tour of the Electronic Computer Project, suffering memory-tube difficulties at the time, and a late-night poker game in the boardroom on the fourth floor of Fuld Hall, the correspondent attended a dance in the common room on the ground floor.

"The participants were mainly Oppenheimer's young atomic physicists and mathematicians," *The New Yorker* observed. "It was like a square dance at any small college, except that more languages were in use. The boys were in shirtsleeves, slacks or jeans, and sneakers; the girls—some of them wives, some of them secretaries, some of them scientists—were mostly wearing peasant blouses, dirndls, and either saddle shoes or ballet slippers. The music was furnished by a phonograph, and the calling was routine, but every once in a while we overheard one of the young men in a square explaining a step to the rest, and it sounded like '. . . then you form a rhomboid figure with discrete vectors at an angle of 2 pi over M.' One girl, who looked about eighteen, was pointed out to us as a distinguished Old World mathematician called to the Institute after she had published a revolutionary paper on non-associative algebras. Later, she and a distinguished astrophysicist of twenty-two sat down together on a rolled-up rug."[41]

The physicist typing a letter to his mother was Freeman Dyson. The statistician calling the dances was John Tukey, who had recently coined the contraction of "binary digit" to "bit." A bit is the smallest possible unit of information, representing the difference between any two alternatives: a zero or a one, the presence or absence of a hole in a punched card or paper tape, or the difference between a black marble and a white marble in the digital computer Leibniz envisioned in 1679. In seven-bit ASCII code, the abbreviation of "binary digit" to "bit" saves sixty-three bits.

The distinguished astrophysicist was Dick Thomas, age twenty-eight at the time. The mathematician who looked about eighteen was twenty-five. Her name was Verena Haefeli (née Huber), and in 1947 she had received her Ph.D. from the University of Zurich with a dissertation in finite group theory titled "Ein Dualismus als Klassifikationsprinzip in der abstrakten Gruppentheorie" (A dualism as a classification principle in abstract group theory), leading to her invitation to the IAS. She was born in Naples to Swiss parents in 1923, moving to Athens in 1929, where her father, Charles Huber, managed the Middle Eastern operations of Bühler AG, a Swiss food-process engineering firm, while she attended the German-speaking school, took shop classes with the boys, and developed a growing interest in math.

Her leanings toward engineering were discouraged as unconventional, and she turned to mathematics as a more acceptable escape. She could observe the requisite social appearances outwardly while playing with numbers in her head. "When I was eight or nine we were doing linear equations with a new Math teacher," she remembers. "When I asked him 'how about the equation $x^2 = 10$,' he patted me on the head and said 'you have got to be patient, we'll get there in good time.' So I went home, settled myself on the couch in our living room and first of all got a clean proof of the fact that the square of no rational number m/n (m and n ordinary integers) could possibly be equal to 10 and then proceeded to construct an infinite sequence $\{m_i/n_i\}$, of rationals whose square 'converged' to 10."

She was too shy to show the result to her math teacher. "So it remained a secret between me and myself."[42]

During the 1930s the Deutsche Schule Athen began to change. "Pretty soon the teachers the Reich sent us no longer had Ph.D.'s," she continues, while some of her favorite teachers, such as Dr. Preibisch and Dr. Lichtenstein, began to disappear. Her German classmates began joining the Hitler Youth. She and her younger sister Heidi found "a couple of Swiss Alpine Club pins, consisting of a big white Swiss Cross in its red square field surrounded by a few Edelweiss, an ice axe and a pair of crampons," that "we pinned to our blouses and whenever we were hailed by a brisk Teutonic 'Heil Hitler' we would point with our right index finger to the SAC pin over our heart and say quietly but with determination 'Heil Dir Helvetia.'" (Hail Switzerland.)

Verena, Heidi, and their mother moved back to Zurich in 1940 while their father wound up his business in Athens and joined the International Committee of the Red Cross (ICRC). He was deployed to India and Ceylon, monitoring prisoners of war being held in British camps. Telegrams arrived sporadically, via International Red Cross headquarters in Geneva, because wartime censorship was in effect. Switzerland was spared the physical destruction of the war, while Charles Huber bore witness to its psychological effects. Red Cross officials were one of the only channels for bringing outside mail into the camps. The Red Cross visits were "a rare ray of light for the prisoners," says Paolo Mastrolilli, whose father, an officer in the Italian army, was wounded in Libya and sent to a British camp in India in 1943. "I think what Mr. Huber did was precious to many people and most likely saved many lives."[43]

Verena was denied admission to ETH (the Eidgenössische Technische Hochschule, or Swiss Federal Technical Institute) because she was not yet eighteen, so she enrolled in the University of Zurich, where she began working under Andreas Speiser, who had published a classic text on finite groups in 1923, the year she had been born. She proposed a master's thesis on the structure of finite groups, to which Speiser, a student of David Hilbert's, responded by advising her to submit the thesis for a Ph.D. instead.

She had been drawn to group theory at first sight. "I could *see* the transformations," she explains. "In nature you find groups wherever there are symmetric shapes or events." Abstract group theory, brought into modern form by the twenty-year-old and already twice-imprisoned French mathematician and political agitator Évariste Galois in a frantic last-minute night's work before being mortally wounded in a pistol duel on the morning of May 30, 1832, has been described, in a definition that is broad "but not trivial," as "a branch of mathematics in which one does something to something and then compares the result with the result obtained from doing the same thing to something else, or something else to the same thing."[44]

My mother, Verena Huber-Dyson, used to explain mathematics to her young children by asking us to imagine a shepherd whose flock leaves in the morning and returns at night. The shepherd collects a set of pebbles and, by adding a pebble as each sheep leaves and subtracting a pebble as each sheep returns, maps the pebbles to the sheep and thus keeps track of her flock. While waiting around, she might start playing with the pebbles, forgetting about the sheep, and notice that there are not only different sets of pebbles, and even sets of sets of pebbles, but also different possible transformations and symmetries between sets of pebbles. Eventually, forgetting about the sets of pebbles as well as the sheep, and just playing around with the transformations and symmetries, she might discover that they fall into structures, the way stars form constellations in the sky. And that would be the beginning of group theory.

To obtain a degree from the University of Zurich, group theory was not enough. Verena chose chemistry as her minor subject, believing that all chemistry could be understood "as an algorithm based on the science of atomic physics," whose details could be derived as needed from there. She showed up for her chemistry exam on a spring morning at the end of the war in Europe in 1945. "All I could see at the far end of the room was the large and familiar expanse of the back side of the morning edition of the

Neue Zürcher Zeitung unfolded and upheld, with a couple of well-polished shoes sticking out from under its lower margin. After the *NZZ* had slowly, deliberately and neatly been folded and laid on the table by Professor Karrer [Paul Karrer, author of the classic textbook *Organic Chemistry*] the disaster struck. He asked me to explain how gunpowder is made." She failed and had to take chemistry again.[45]

Verena's mother, Berthy Ryffel, died of heart failure in June 1945, and her father was killed in a nighttime collision while returning in an ambulance from a hospital inspection in the British sector of Germany in November 1946. In the year before his death he had made an extended tour of American prisoner-of-war camps as an ICRC delegate to the United States, which he described to Verena as a place she "definitely ought to experience at length and in depth but just as definitely ought not to settle in." After receiving her Ph.D., Verena sailed for America aboard the Holland America liner *Nieuw Amsterdam* over New Year's of 1947–1948 with some money from her father's estate, her two-year-old daughter, Katarina, and just enough English to get by when she arrived in the United States.

Her first four months in the United States were spent at the University of Illinois at Urbana under the auspices of Reinhold and Marianne Baer, who helped her adjust to life in America and enlisted Solomon Lefschetz at Princeton to facilitate her invitation to the IAS. Princeton University would not appoint a woman professor until later in 1948, and female undergraduates would not be admitted until 1969. The institute offered sanctuary not only to refugees fleeing the Nazis but also to women who otherwise were denied entrance to the Princeton academic community at that time.

Particle physics took the lead under Oppenheimer's tenure at IAS, with group theory playing a supporting part. "Oppenheimer had a very narrow view; he thought that anything that was not fundamental science shouldn't be done," Freeman Dyson remembers. "And for him, the only thing that was really fundamental was particle physics."[46] As postwar cyclotrons and linear accelerators came online, a cascade of new elementary particles started turning up. The puzzle was that when one tried to understand these most fundamental things, it became clear that what really exists is not *things* but *fields*. Field theory could explain the behavior of these particles but failed to explain *why* these particular families of particles, and not others, were being found. It was group theory, with its emphasis on symmetries, or things that don't change when something else changes, that held the key to understanding the genealogy of elementary particles and, in consequence, how it is that our world of material substance, from pebbles to sheep, emerges from nothing except fields.

Thus my mother and father found themselves at the Institute for Advanced Study for the fall term of 1948. Freeman spent much of his time taking long walks by himself. "Unfortunately my young colleagues are unwilling to join me," he complained, "as they are obsessed with the American idea that you have to work from 9 to 5 even when the work is theoretical physics. To avoid appearing too superior, I have to say that it is because of bad eyes that I do not work in the afternoons."[47] At the end of the term, he received an unusual offer from Oppenheimer: a five-year appointment to the School of Mathematics, whose obligations could be fulfilled by being in residence as much or as little as he wished. He accepted, although expressing to Oppenheimer "a feeling of doubt whether it is good for the character of any man at my time of life to be paid so much for doing nothing."[48] Verena was invited to stay, without a stipend, for a second term that would end in May 1949. They were about to part ways at the time of the *New Yorker* visit in April 1949.

As a single mathematician, Freeman lived in a rented room in downtown Princeton, taking meals either at the institute cafeteria or among the inexpensive eateries in town. "The Housing Project, where all the young married people with their families live," he wrote to his mother, "has seemed to me until very recently a Garden of Eden from which I was excluded by an unkind fate." On April 7, 1949, everything changed. "I was strolling around by myself, taking the air and feeling somewhat melancholy, and I happened to pass Verena Haefeli's house on the housing project, and little Katrina [*sic*] was standing at the window and began shouting at me as I went by," he continued. "This was odd, because I had hardly spoken to her before to-day."[49]

Katarina ushered Freeman inside, where he spent the rest of the afternoon learning about Verena's life and work. She not only had a Ph.D., a daughter, and a recent divorce; she also had a car: a maroon 1940 Dodge convertible coupe with a 217-cubic-inch straight six. The next day they drove the Dodge, with the top down, to the beach. They took repeated drives to the New Jersey shore during the time they had left together in Princeton and made a joint appearance before a justice of the peace in Lakeside, New Jersey, on May 15, 1949, to settle a ten-dollar fine resulting from Verena's running a stop sign the previous weekend on the return trip.

They found that they shared, among other things, an interest in Kurt Gödel's work on the foundations of mathematics: not only his incompleteness theorems of 1930 and 1931, but also the work he had been engaged in since taking permanent refuge at the institute after leaving Vienna for the last time in 1939. Freeman presented Verena with an annotated copy of Gödel's monograph *The Consistency of the Continuum Hypothesis*, concerning the unresolved question of whether infinities can ever be found in anything other than two sizes: uncountably large or countably small. Freeman's copy contained six additional pages of penciled notes demonstrating that "by making certain alterations in the arguments of this tract . . . we have therefore proved a stronger result than that stated in the text."[50]

Freeman and Verena were married before a justice of the peace in Michigan in August 1950, before taking a honeymoon tour of the Upper Peninsula by car. Their first child, Esther, was born in Zurich in July 1951, and in 1953, back in Ithaca, New York, I followed. Oppenheimer then offered Freeman a permanent appointment to the institute, where he has remained, with only occasional sabbaticals, ever since.

After dinners in the cafeteria on the fourth floor of Fuld Hall, Esther and I would return our trays to the kitchen before sneaking off to explore the corridors and stairwells where unaccompanied children were not allowed.

Esther could read the clues that identified the offices, but I could only distinguish between the wing belonging to the historians and the wing belonging to the mathematicians by the smell. Perhaps the mathematicians, thinking more intensely, left traces of perspiration in the air. Or perhaps it was just that they used chalkboards and favored pipe tobacco, whereas the historians favored bookshelves and smoked cigars. These distinctions were sensed unconsciously, the way a young animal recognizes its own den.

To its scholars, the institute was at the center of academia. To us children, it was in the middle of the woods. Oswald Veblen, whose family had lost its land in Norway to unscrupulous lawyers before carving out a series of homesteads on the American frontier, was obsessed with acquiring land. He persuaded the Bambergers to underwrite the purchase of several adjacent holdings in addition to Olden Farm, which had changed hands only twice since its acquisition by William Penn. Before that it had belonged to the Leni Lenapes, the original inhabitants of New Jersey and Pennsylvania, whose language and customs Penn studied by traveling among them before settlements were formed. "I find them . . . of a deep natural sagacity," he wrote to his friend Robert Boyle of the Royal Society in 1683. "The low dispensation of the poor Indian out shines the lives of those Christians, that pretend an higher."[51]

The woods were bounded by the institute grounds and housing project on one side, by the original 1726 Friends Meeting House and 1777 Princeton Battlefield on another, by the Princeton Junction rail line and Elizabethtown waterworks on yet another side, and by the Delaware and Raritan Canal, with a shantytown along its banks whose inhabitants, the descendants of freed slaves, had once worked the fields. In the middle of the woods, among a grove of old-growth beech trees and without electricity or running water, lived a single family of tenant farmers, left from when the institute had purchased the land.

Veblen always dressed for the outdoors, kept an ax for clearing windfalls in his office, and played Christopher Robin to the institute's six-hundred-acre woods. The boundaries were blurred, at bedtime, between the woods surrounding the institute and the map of the Hundred Acre Wood that was printed on the endpapers of *Winnie-the-Pooh*. Christopher Robin's Hundred Acre Wood was inhabited by a bear, a piglet, an owl, a donkey, a tiger, a rabbit, and two kangaroos. I belonged to a small band of eight-year-olds who were more interested in reptiles than mammals or birds. We hunted snakes, turtles, and frogs: capturing them, then setting them free. We avoided water moccasins, a subspecies of rattlesnake, and snapping turtles larger than dinner-plate size. The snapping turtle (*Chelydra serpentina*) is a survivor of the Cretaceous that has remained unchanged for sixty million years Now limited to eastern North America, they once ranged across a warmer and wetter world. It was as if dinosaurs were still roaming the New Jersey woods.

Reaching up to sixty pounds at maturity, they exhibit uncharacteristic agility for a turtle, with an elongated neck terminating in massive jaws that can crush a stick as thick as a child's arm. Adults have no natural predators and can live for more than a hundred years. They are powerful swimmers and almost impossible to catch in the water, burrowing into the sediment and spending most of the time submerged. Occasionally a snapping turtle will go lumbering through the woods like a heavily armored tank to find a mate, lay its eggs, or move to another body of water, and can be captured if unable to reach water to escape.

When we were about the age of ten, our interests shifted from reptiles to machines. Near the edge of the woods, at the lower end of Olden Lane, was a barn left from the days of Olden Farm. It still contained bales of hay, farm equipment, and other relics of its previous life. A row of former milking stalls had been appropriated by the Electronic Computer Project and contained piles of war-surplus electronic equipment that had been cannibalized for vacuum tubes and other parts.

We slipped through a gap in the siding and picked over these remains. As children of the 1950s, we had been indoctrinated with a fear of vacuum tubes, especially cathode-ray tubes, which operate at voltages that could kill curious children who stuck their sweaty hands into the cabinet of a TV. So we extracted relays, solenoids, microswitches, and other low-voltage electromechanical components that we could connect to battery or doorbell transformer power at home. At the edge of the safe-voltage envelope were large dry-cell batteries we acquired from fathers who worked for RCA. Portable battery-powered vacuum tube equipment took three different battery voltages to power up the tubes. The snapping turtles of dry-cell batteries were milk-carton-sized B batteries used to supply the plate voltages: delivering a forty-five-volt and ninety-volt kick.

Life-forms, technologies, and institutions often enter a period of extravagance before going extinct. The age of the transistor was coming, and for vacuum tubes the time was running out. This was the era of tail fins, menthol cigarettes, and an obsession with vacuum-tube-driven hi-fi audio equipment that filled sofa-sized cabinets and delivered a lossless frequency spectrum wider than that detectable to the human ear. One of the reasons the otherwise unwelcome Electronic Computer Project was tolerated at the IAS, after the hydrogen bomb calculations were completed, was that the project engineers kept most of the key faculty, including Oppenheimer and Einstein, supplied with hi-fi systems better than anything obtainable on the commercial market at the time.

Vacuum tubes were too successful for their own good. People wanted too many of them, they were too hard to make, their working lifetime was too short, and it took too much electricity to heat their filaments to a temperature at which electrons would venture out into the vacuum and do their work. On December 23, 1947, at the Bell Telephone Laboratories in Murray Hill, some thirty-five miles from Princeton, the first crude transistor was brought to life. The era of the vacuum tube was drawing to a close.

One by one, then by the thousands, and soon by the millions, triodes became transistors. Instead of blown-glass envelopes containing individually crafted electrodes that had to be heated red-hot to coax electrons into a state where they would do what they were told, the transistor achieved the same results at room temperature and low voltage, with monolithic silicon enabling manufacture at ever smaller size and ever larger scale. Vacuum tubes vanished without a trace, except among a small circle of audiophiles still seeking the warmth they bring to a room.

Snapping turtles, in contrast, are still emerging out of the Cretaceous, one pond at a time, in the New Jersey woods. A warmer, wetter world will suit them just fine. The Anthropocene, so far, is but the blink of a double-lidded eye in their time upon the earth.

4

VOICE OF THE DOLPHINS

I n September 1933, Leo Szilard, a thirty-five-year-old Hungarian physicist, was wandering around London, knocking on doors "trying to find positions for German colleagues who lost their university positions with the advent of the Nazi regime."[1] Szilard was a friend, student, and colleague of Albert Einstein's and, along with John von Neumann, Eugene Wigner, Theodore von Kármán, and Edward Teller, one of the five Hungarian "Martians" whose immigration to America in the 1930s sparked the development of nuclear weapons, digital computers, and the intercontinental ballistic missile, promoted by von Neumann as "nuclear weapons in their expected most vicious form."[2]

Szilard obtained his Ph.D. from the University of Berlin in 1922 with a dissertation on thermodynamics that earned Einstein's enthusiastic approval and led, within six months, to a landmark ten-page paper, "On the Decrease of Entropy in a Thermodynamic System by the Intervention of Intelligent Beings," published in German in 1929 and in English only after Szilard's death in 1964. Szilard invoked a connection between entropy and

View of the Land to the N.º of Foggy Island A bearing S.W.½W. lea. dist. and B.W.⅓ 4 lea.²dist.

information to resolve the question of whether Maxwell's demon, an intelligent agent "who catches the fast molecules and lets the slow ones pass," can, by opening and closing a doorway into a closed compartment in a randomly fluctuating thermodynamic system, raise its temperature without the expenditure of physical work, thereby beating the house.[3]

Szilard distilled the problem down to a single particle confined to a box divided into two parts. What is the lowest possible cost of knowing which side of the box contains the particle at any given time? He analyzed the thermodynamic consequences of this minimal physical representation of what we now term one "bit" of information, but it would be another two decades until the current terminology took hold. Szilard's insights, along with those of the communication theorists Harry Nyquist and Ralph Hartley, influenced John von Neumann and Norbert Wiener, anticipating Claude Shannon's formulation of information theory in 1948. They also led to a collaboration with Einstein that resulted in a patent for a refrigerator with no moving parts.

Szilard reduced Maxwell's demon, a vaguely defined being "whose faculties are so sharpened that he can follow every molecule in its course,"[4] to a device whose only intelligence was the ability to store, recall, and forget one bit of information at a time. This "seems to us to be the essential thing," he observed, in seeking to "determine whether in fact a compensation of the entropy decrease takes place as a result of the intervention by such a device."[5] Long before the age of digital computers, Szilard addressed their ultimate constraint: What are the limits to the cost of remembering—and forgetting—all these bits of information from one instant to the next?

Even if you make the individual marbles in Leibniz's computer vanishingly small, knowing where they are invokes a certain non-negligible expense. In a machine processing billions of bits billions of times per second, the cost adds up. Analog computers, in contrast, do not require tracking all the marbles at every increment of time.

———

During Szilard's twenty-six years in the United States, he never held a conventional job, owned no more than would fit in a suitcase, lived mostly out of hotel rooms, and was trailed persistently by the FBI. His life was a series of contradictions. He filed dozens of patents, ranging from the Einstein-Szilard refrigerator to nuclear reactors to improvements in methods for the preparation of caffeinated beverages, and even patented (under British Admiralty secrecy) the prospect of a nuclear chain reaction in 1934. Yet he never reaped any significant rewards. He admired the wisdom and honesty of children but never had a family himself. After helping to bring nuclear weapons into existence, he campaigned against them for the rest of his life.

Szilard was annoyed, on that September day in London, not only at the treatment of his fellow Jewish scientists by the German government but also by a recent "warning . . . to those who look for sources of power in atomic transmutations [that] such expectations are the merest moonshine," delivered by Lord Ernest Rutherford, the nuclear pioneer.[6] "Pronouncements of experts to the effect that something cannot be done have always irritated me," Szilard recalled. "That day as I was walking down Southampton Row and was stopped for a traffic light [at Russell Square], I was pondering whether Lord Rutherford might not prove to be wrong. As the light changed to green and I crossed the street, it suddenly occurred to me that if we could find an element which is split by neutrons and which would emit *two* neutrons when it absorbed *one* neutron, such an element, if assembled in sufficiently large mass, could sustain a nuclear chain reaction, liberate energy on an industrial scale, and construct atomic bombs."[7]

Szilard had been introduced to nuclear weapons by H. G. Wells, whose novel *The World Set Free* appeared in 1914 on the eve of World War I. Wells chronicled a landscape transformed by atomic energy and destroyed by atomic bombs, before an enlightened world government emerges from the ashes of global war. Szilard, who met Wells in London in 1929, subscribed to his arguments for international control of atomic energy and embraced his prediction that "a day will come, one day in the unending succession of days, when beings, beings who are now latent in our thoughts and hidden in our loins, shall stand upon this earth as one stands upon a footstool, and shall laugh and reach out their hands amid the stars."[8]

World War I, "the war that will end war," as Wells labeled it,[9] came to an end without nuclear weapons, and when Szilard took up nuclear physics in the 1920s, he was envisioning space exploration, not bombs. "Only through the liberation of atomic energy could we obtain the means which would enable man not only to leave the earth but to leave the solar system," he remembers thinking at the time.[10]

Szilard arrived in America on the recommendation of Einstein in 1938. On August 2, 1939, he drafted a letter to President Roosevelt, signed by Einstein, warning of the possibility of the atomic bomb. Under persistent badgering by Szilard, the U.S. government promised six thousand dollars for "uranium research," but it was March 1940 before the money was released. Impatient at the delay, Szilard rented some thorium himself, seeking private investors to underwrite the construction of a nuclear reactor in exchange for shares. This entrepreneurship alienated his academic colleagues, failed to attract investors, and was viewed as insubordination by the government officials who were plodding their way through the proper channels to secure funds.

Enrico Fermi and Leo Szilard were the first two scientists to join the U.S. Army's Manhattan Engineer District, as the atomic bomb project was code-named. The welcome, in Szilard's case, was short. On August 13, 1940, the army was advised, on the basis of information "received from highly reliable sources" and forwarded to J. Edgar Hoover at the FBI, that "employment of this person on secret work is not recommended," with relations deteriorating from there.[11] General Leslie Groves, who had supervised the construction of the Pentagon and was given command of the Manhattan Project, took such offense at Szilard's disrespect for military authority that he excommunicated him from Los Alamos and attempted to have him interned as a dangerous enemy alien for the duration of the war.

"Szilard is loyal to the United States, but was a source of considerable difficulty to the War Department," an FBI summary explains. "Even before the atomic bomb was completely developed, he began a concerted campaign to utilize the developments of the Project to promote lasting peace after the war. Although he was one of the first to advocate secrecy of the Project, he had his own ideas how secrecy should be handled, and they differed from the ideas of M.E.D."[12]

When the surrender of Germany ended the threat against which the Manhattan Project had been launched, Szilard began lobbying to prevent the use of nuclear weapons against the civilian population of Japan. On July 17, 1945, the day after the Trinity explosion in New Mexico, he submitted a petition to President Truman, signed by sixty-seven of his colleagues in Chicago, asking the president to reveal the existence of the weapon and offer the Japanese an opportunity to surrender, followed by a demonstration against an unpopulated target, if needed, first.

Oppenheimer refused to allow Szilard's petition to circulate at Los Alamos. General Groves delayed its delivery to Secretary of War Henry L. Stimson, preventing its reaching the president until after Hiroshima had been destroyed, and ordered the text of the petition and even any reference to its existence kept secret on the grounds of national security, with a stern warning from the army intelligence office delivered to Szilard. Edward Teller, the Hungarian physicist who was already working on the thousand-times-more-powerful hydrogen bomb, sympathized with Szilard's sentiments but refused to sign the document, answering, "I have no hope of clearing my conscience. The things we are working on are so terrible that no amount of protesting or fiddling with politics will save our souls."[13]

After the destruction of Hiroshima and Nagasaki, the physicists who had created these weapons divided into three camps: those who wanted to build and test as many new bombs as possible, those who wanted to stop building bombs altogether, and those who wanted to keep building bombs but use them for something else.

Leading the first group were the weaponeers at Los Alamos and later Livermore National Laboratory, whose rivalries became as much a driver of the nuclear arms race as the standoff between the United States and the U.S.S.R. "They [Los Alamos] would try to build smaller and better bombs and we would try to build even smaller and even better ones," says Morris Scharff, a physicist who played for the Livermore side.[14]

Leading the second group was a series of arms control campaigns, beginning with the *Bulletin of the Atomic Scientists*, founded in 1945, and the Emergency Committee of Atomic Scientists convened by Einstein and Szilard in 1946.

Taking the lead on the third front was Project Plowshare, a U.S. effort, mirrored by a Soviet counterpart, to use underground nuclear explosions to liberate oil and gas reserves, excavate artificial harbors, and dig canals. There was also a commercial entry into the field: General Atomic, a division of the General Dynamics Corporation established in San Diego in 1955 to commercialize nuclear energy in all forms. The founder was Frederic "Freddy" de Hoffmann, a young Viennese protégé of Edward Teller's who had performed the initial calculations for what was then known as the "Super" bomb.

"I wanted to do something about the hydrogen bomb and nobody else wanted to," remembers Teller, speaking of the period before the successful Soviet bomb test in 1949 prompted President Truman to push American H-bomb development full speed ahead. "And the one man who wanted to do it more than I was Freddy de Hoffmann."[15] It was de Hoffmann, twenty years old at the time, who had calculated the ballistic trajectories for Fat Man and Little Boy, the atomic bombs that were dropped on Japan. "Making the bomb tables for Hiroshima and Nagasaki is etched so strongly in my mind because it really brought me to the reality of the end use of our scientific and technical experimentation," he recalled.[16]

General Atomic operated out of a converted elementary school in Point Loma for its first two years before moving to a futuristic headquarters constructed on an empty mesa north of La Jolla, on the outskirts of San Diego, at Torrey Pines. In 1943, Robert Oppenheimer and the U.S.

Army had requisitioned the Los Alamos Ranch School to begin assembling the physicists needed to build an atomic bomb. Thirteen years later, Freddy de Hoffmann and General Atomic took over the Barnard Street School, vacant since the exodus of military families from San Diego after the war. Many of the same physicists, along with newcomers such as Freeman Dyson, showed up. "The drinking fountains were down very low for children and the blackboards were low," remembers Brian Dunne, an experimentalist who joined the schoolhouse gang. "The machine shop used to be the kindergarten and all the drawers were way down there next to the floor."[17]

General Atomic's first product was an intrinsically safe nuclear reactor, designed to shut down in milliseconds, without human or mechanical intervention, in the event that cooling failed. The goal, as Edward Teller put it, was a reactor that was "not only idiot-proof, but Ph.D.-proof." Those who gathered at the schoolhouse for the summer of 1956 recaptured the enthusiasm of wartime Los Alamos, transgressing the boundaries between physics and engineering to advance from a theory of the warm neutron effect to a two-thousand-megawatt prototype in under two years. When the second United Nations "Atoms for Peace" conference convened in Geneva in September 1958, General Atomic showed up with a working reactor and stole the show. "Everyone wanted to see the blue light," says Dunne. "They sold those things like hotcakes."[18]

More than sixty-five copies of the TRIGA (Training, Research, Isotopes, General Atomic) reactor were commissioned, some still in operation today. TRIGA was consistently profitable, a record unmatched by any other nonmilitary reactor design. At the end of the first summer in the schoolhouse, Freeman Dyson returned to Princeton, leaving his name, along with Andrew W. McReynolds and Theodore B. Taylor, on a patent for a "Reactor with Prompt Negative Temperature Coefficient and Fuel Element Therefor." They received one dollar each for their rights. "The primary object of the present invention is to provide an improved neutronic reactor which will not be destroyed even if grossly mishandled," they explained.[19]

General Atomic believed the reactor's safety needed to be demonstrated in a spectacular way. At the public dedication, with Niels Bohr in attendance, "we pulled out all the control rods explosively so that the reactor was prompt super-critical with a neutron doubling time of two milliseconds," Dyson remembers. "That was the worst accident that we could imagine. The reactor quietly shut itself down in a few milliseconds with no damage. We then repeated the process five hundred times."[20]

Theodore "Ted" Taylor, the leader of the TRIGA project, was a gifted young physicist from Los Alamos who had designed the most powerful fission weapons ever to populate the U.S. stockpile before he obtained his Ph.D. He had grown up in Mexico City, where his father was general secretary of the YMCA. His mother, the daughter of Congregational missionaries, gave him a chemistry set at age seven, with one restriction, as he remembers it: "Never, never under any circumstances was I allowed to make nitroglycerine." So he made nitrogen iodide, picric acid, and other high explosives instead, and as he grew older and found himself "at loose ends" after school, he prowled the streets of Mexico City, looking for places "where the billiard balls were really spherical and the tables were very heavy and flat."[21]

After joining the navy during World War II, Taylor obtained a degree in physics from Caltech through the navy's V-12 scholarship program before heading to Berkeley for a Ph.D. He was forced to withdraw after failing his preliminary examinations, twice. Newly married and with a child on the way, he thought his luck had run out. In February 1949, thanks to Carson Mark, the Canadian physicist who had succeeded Hans Bethe as director of T-Division at Los Alamos, he was offered a job working on "problems in neutron diffusion," paying $375 a month.

Taylor had been at midshipman's school in Fort Schuyler, New York, when reports of the Hiroshima bombing came in. "I didn't know what was going to happen to me," he remembers, "but I did know one thing, and that was that I would never work on nuclear weapons. Four years later

I was not only working on nuclear weapons at Los Alamos, but doing so with considerable enthusiasm. It turned out I was really good at it."[22] His understanding of nuclear explosions, and how to create the conditions it took to initiate them, manifested itself in an uncanny, intuitive way. It was a game of billiards played with atomic nuclei, and at Los Alamos the tables were perfectly flat.

"I had complete freedom to work on any new weapon concept I chose," he says. "It's an exhilarating experience to look at what's going on theoretically, on paper, inside something the size of a baseball that has the same amount of energy as a pile of high explosive as big as the White House. I went crazy over that. A big high. The highs needed fixes. And we got those twice a year easily. The fix was a combination of seeing one of these things go off—'Aha! It worked!'—and seeing how the next one might be even more spectacular."[23]

Taylor was soon designing bombs whose yields were larger than had been thought possible while making them smaller and more efficient as well. "I was playing around with the middle of implosions, the last millimeter or two, and Carson [Mark] said, 'Keep your eyes open for high temperatures,'" he remembers. "Because there's always a possibility that you could put some deuterium inside, and get neutrons out. Which had all kinds of implications." The technique became known as boosting and remains the state of the art. "Oppenheimer was very scornful," Taylor explains. "He used to go around and ask people what they were doing when he was on the scientific advisory committee for the AEC, and he said [that boosting] is a waste of time, because of the Taylor [hydrodynamic] instability. But he was just dead wrong."[24]

The largest fission weapon that Taylor designed was the Super Oralloy Bomb (SOB), yielding five hundred kilotons in the Ivy King test at Eniwetok on November 15, 1952. His goal was a fission explosion so powerful there would be no military targets left for the hydrogen bomb. Some ninety copies of the SOB were deployed as a stockpile warhead, but Taylor's strategy failed. Hydrogen bomb production, aimed at civilian populations, was nonetheless pushed ahead.

Taylor then shifted his attention to the opposite end of the scale. "Why can't we push that limit," he asked, "getting the quantities of plutonium to make nuclear explosions down into less than a kilogram? Quite a bit less. I tried to find out what was the smallest bomb you could produce. It was a full implosion bomb that you could hold in one hand that was about six inches in diameter."[25]

In 1953, Taylor was sent on paid leave to Cornell University to obtain his elusive Ph.D. It had become a growing embarrassment to have the design of the nation's most advanced weapons in the hands of someone without an advanced degree. Taylor's designs continued to be built and tested while he was a graduate student, under Hans Bethe's supervision, at Cornell. He remembers it being "unbearably exciting" to receive an occasional cryptic phone call from Carson Mark at Los Alamos: "Well, how did the wasp go?—Just great!"[26]

Dyson and Taylor, both protégés of Bethe's, became close friends during the summer at General Atomic in 1956. Taylor was as well liked by the machinists on the shop floor at General Atomic as he was by the generals at the Pentagon. "He didn't play big shot," says Jaromir Astl, a technician who transferred to General Atomic from Convair, "he played one of the guys."[27] At the end of the summer, Taylor accepted an offer to remain at General Atomic, finding life in La Jolla more suited to his growing family than Los Alamos. He was ready to settle down to designing reactors instead of bombs. Then *Sputnik* went up.

On October 4, 1957, the Soviet Union launched an artificial satellite into orbit around the earth. What could the United States launch in response? "That night," Taylor recalls, "I derived for myself the notion that if you really added together the features that you wanted of any vehicles for exploring the solar system—the whole thing, not just near in—you're led directly to energy on the scale of a lot of nuclear weapons. And having been led to that, in thinking about what might be done, I began saying, 'Gee, that's what Stan Ulam's been talking about all these years.'"[28]

The Polish mathematician Stanislaw Ulam, one of Taylor's mentors at Los Alamos, having observed that parts of the Trinity shot tower, made of ordinary steel, had not been vaporized by the fireball and had survived the explosion intact, began asking, Why not use nuclear explosions to propel rockets rather than using rockets to deliver bombs? He pushed the idea as far as to take out a patent, with Cornelius J. Everett, for a "nuclear propelled vehicle, such as a rocket,"[29] and wrote a Los Alamos report, "On a Method of Propulsion of Projectiles by Means of External Nuclear Explosions," issued (in secret) in August 1955.

"Repeated nuclear explosions outside the body of a projectile are considered as providing a means to accelerate such objects to velocities of the order of 10^6 cm/sec . . . in the range of the missiles considered for intercontinental warfare and even more perhaps, for escape from the earth's gravitational field," they explained. The energy and temperature limitations governing internal combustion rockets could be raised by several orders of magnitude by taking a nuclear reactor to its extreme—the explosion of a bomb—and isolating the resulting high temperatures from the ship. "The scheme proposed in the present report involves the use of a series of expendable reactors (fission bombs) ejected and detonated at a considerable distance from the vehicle, which liberate the required energy in an external 'motor' consisting essentially of empty space."[30] Orion was a very large external combustion engine. Can you blow something *up* without blowing it apart? The answer appeared to be yes.

"I was up all night, and then I got alarmed that things were getting big," Taylor recalls. "Energy divided by volume is giving pressure, so the pressures were out of sight, unless it was very big. It got easier as it got bigger. I was thinking of something that might carry a couple people, with shock absorbers. I went in to General Atomic the next morning, and my office was right next to Chuck Loomis, who had come down from Los Alamos, and I told him about this sense of discouragement because it was so big. And he said, 'Well, think big! If it isn't big, it's the wrong concept. What's wrong with it being big?' In less than thirty seconds everything flipped. It was Chuck's call that if you were serious about exploring the solar system, why not use something the size of the *Queen Mary?* He understood that bombs could in principle do it. They could lift downtown Chicago into orbit."[31]

To the U.S. military, Sputnik was a catalyst but not a surprise. The Atlas and Titan missiles, the Explorer and Vanguard satellite programs, the Rover nuclear rocket project, and even plans for a moon landing had been under way for a decade or more. Long-standing rivalries stood in the way. "If it *flies*, that's our department," claimed the air force. "But they're called space*ships*," replied the navy. "Okay, but the Moon is high *ground*," answered the army, which was in the lead, having enlisted the German rocket pioneer Wernher von Braun. President Eisenhower established the Advanced Research Projects Agency (then ARPA, now DARPA) to coordinate both civilian and military space programs in the effort to catch up. NASA, requiring an act of Congress, would not exist until July 1958.

General Atomic issued Taylor's *Note on the Possibility of Nuclear Propulsion of a Very Large Vehicle at Greater than Earth Escape Velocities* on November 3, 1957, the day that *Sputnik II*, carrying the dog Laika, was launched. The new project was code-named Orion—for no particular reason, says Taylor, who just "picked the name out of the sky." At the end of November, de Hoffmann went to Princeton to recruit Freeman Dyson, the proposal being too secret to discuss by mail or phone. "He came here and he said, 'Look, you've got to come to GA,'" says Dyson. "I said no, I have no intention of coming to GA, I've done my bit for GA. And he said no, you must come, we have something much bigger and much more exciting, and then he told me Ted had this wonderful scheme for getting around the solar system with bombs."[32]

Dyson was a follower of Jules Verne and H. G. Wells. At the age of eight, he had begun writing a sequel to Verne's *From the Earth to the Moon and a Trip Around It*, based on a predicted collision between the asteroid Eros and the moon. *Sir Phillip Roberts's Erolunar Collision* was left unfinished when Dyson turned nine. While Taylor had been detonating small charges of homemade explosives under passing streetcars in Mexico City, Dyson had been theorizing about using nitrocellulose to propel a fifteen-foot-diameter spaceship to the moon. He first estimated how large a moon-based launcher it would take to escape from the moon's gravitational field for the return to earth. He then estimated how large a terrestrial launcher it would take to deliver the lunar launcher to the moon. "When I thought about space-travel in those days, I was thinking about the huge guns that I read about in the stories of Jules Verne," he explains. "Rockets had nothing to do with it. The Martians in Wells's *War of the Worlds* did not come in rockets. They came in artillery shells."[33]

On New Year's Day 1958, Dyson flew to San Diego to consult with Taylor for ten days. The resulting proposal, submitted to ARPA in early 1958, "included all the necessary practical working features for a very large space vehicle" weighing four thousand tons, carrying twenty-six hundred bombs, and capable of orbiting a payload of sixteen hundred tons.[34] An air force reviewer noted that "the uses for ORION appeared as limitless as space itself."[35] Three weeks later, Stan Ulam testified in Washington before the Subcommittee on Outer Space Propulsion of the Joint Committee on Atomic Energy, and although he could not divulge any details to the assembled congressmen, he hinted that an idea was under consideration that "is almost like Jules Verne's idea of shooting a rocket to the Moon."[36]

Upon his appointment as technical director of ARPA in March 1958, Eisenhower's adviser Herbert F. York found a proposal for a four-thousand-ton interplanetary spaceship on his desk. "It was the time after Sputnik when everybody was looking for some kind of an answer and thinking that technology was the likeliest place to look, and so a lot of stuff that would be too far out under ordinary circumstances managed to get included inside the envelope," he explains.[37]

Long before Sputnik, York had helped set the United States on the path to space, and after Sputnik he produced a classified report for President Eisenhower that laid out plans for a series of large rocket boosters leading to a moon landing within twenty years. York was skeptical of estimated costs. "It wasn't just Orion," he says. "Almost every cost estimate made by a physicist is wildly wrong, and the better the physicist, the worse it is."[38]

York authorized a one-year feasibility study, with "the verbal under-standing that the contract would be extended at a somewhat higher rate if it proved technically impossible to disprove feasibility at the end of the first year."[39] He believed that Orion, however improbable, deserved a chance. "It was a unique time. When we were getting ARPA started, we were willing to take some fliers. I never thought it was feasible, but that's okay, I thought it was interesting. And of sufficiently dramatic ultimate potential that even very low feasibility merited some attention. I tried to put that combination together somehow and multiply that out."[40]

ARPA's contract for the "Feasibility Study of a Nuclear Bomb Pro-pelled Space Vehicle" was issued on June 30, 1958, between the General Atomic division of General Dynamics Corporation and the U.S. Air Force Air Research and Development Command. "If the concept is feasible, it may be possible to propel a vehicle weighing several thousand tons to ve-locities several times earth escape velocities," the contract noted. "Such a vehicle would represent a major advance."[41]

The amount awarded was $949,550 plus a fixed fee of $50,200 for a total of $999,750. "There must have been a million-dollar limit," says Ed Giller, a young physicist and air force colonel who was assigned manage-ment of the contract on the air force side. "Right up against the peg!" adds his colleague Don Prickett.[42]

Ever since the Alamogordo (Trinity) test in 1945 and the Bikini (Cross-roads) tests in 1946, nuclear testing had been expanding from year to year. Bomb tests answered the question of whether a given design would ex-plode, and weapons effects tests answered the question of what happened next. The tests were divided between the Nevada Test Site for small ex-plosions and the South Pacific for large explosions, with San Diego's air and naval facilities falling in between. De Hoffmann, searching for a home where physicists could settle down in the aftermath of Los Alamos,

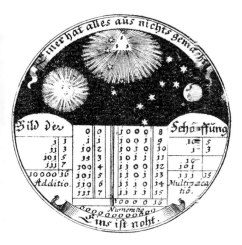

Design for a silver medallion, presented to Rudolph August, Duke of Brunswick, by Gottfried Wilhelm Leibniz on January 2, 1697, illustrating "the creation of all things out of nothing through God's omnipotence." Binary arithmetic, believed Leibniz, held powers extending beyond "those who sell oil or sardines." (After *Herrn von Leibniz' Rechnung mit Null und Eins*, ed. Erich Hochstetter and Hermann-Josef Greve, Berlin, 1966)

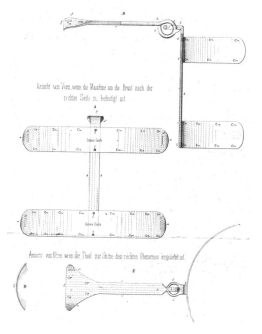

Mechanical support for the paralyzed arm of the tsar, which is to be made by Leibniz. Peter the Great, suffering paralysis in one arm, enlisted Leibniz to design and build a prosthetic support. (After Vladimir Guerrier, *Leibniz in seinen Beziehungen zu Russland und Peter dem Grossen*, St. Petersburg, 1873)

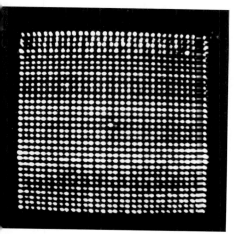

Digital universe in 1953. A 32-by-32 (1 kilobit) array of charged spots—serving as working memory, not display—is visible on the face of a Williams cathode-ray storage tube in this diagnostic photograph from the maintenance logs of the Institute for Advanced Study's Electronic Computer Project, February 11, 1953. At this time there were 53 kilobytes of high-speed random-access memory on planet earth. (Courtesy of Shelby White and Leon Levy Archives Center, Institute for Advanced Study, ECP Records)

First contact: Aleut kayaker, as observed during the encounter between the Russians on board the *St. Peter* and the inhabitants of the Shumagin Islands on September 9, 1741. (Detail from the chart of the voyage of the *St. Peter* by Sven Waxel and Sofron Khitrov, 1744, in the Archives of the Ministry of Marine, St. Petersburg. 1914; after F. A. Golder, *Bering's Voyages*, 1922)

LEFT The Steller sea cow (*Hydrodamalis gigas*), weighing up to eight thousand pounds, was discovered at Bering Island in 1741 and last seen alive in 1768. "As long as things escape us and perish unknown with our consent, and through our silence," noted Georg Wilhelm Steller, "it is not strange that these things . . . have remained to the present time unknown and unexplored." (Lithograph after a sketch made by Friedrich Plenisner on Bering Island in 1742, from P. S. Pallas, *Zoographia Rosso-Asiatica*, St. Petersburg, 1831. Courtesy of Smithsonian Libraries and the Biodiversity Heritage Library)

RIGHT Petroglyph at Surge Bay, Yakobi Island, possibly depicting the longboat in which eleven men from the *St. Paul* went ashore in search of water on July 18, 1741. The longboat carried four oarsmen on each side, a steersman aft, and two nonrowing passengers, along with two large water casks. (Photograph by Don Douglass, courtesy of Réanne Hemingway-Douglass)

Aleut sea-mammal hunters off Unalaska, 1827. (Lithograph by G. Engelmann after a drawing by Friedrich H. von Kittlitz, from the atlas to Frederic Litke's *Voyage Autour du Monde*, Paris, 1835. Courtesy of Yale Collection of Western Americana, Beinecke Library, Yale University)

Kyril Ermeloff and Ivan Suvorov in whale-intestine rain gear at Nikolski, Umnak Island, in 1909. (Photograph by Waldemar Jochelson, courtesy of Carnegie Institution of Washington)

Naiche (mounted, center) and Geronimo (standing beside Naiche's horse) with their band of Chiricahua Apache warriors, in the Sierra Madre, March 26, 1886. Despite "the constant hounding of the campaign," General Crook found the fugitives "in superb physical condition, armed to the teeth, and with an abundance of ammunition." (Photograph by Camillus S. Fly, courtesy of Palace of the Governors Photo Archives, NMHM/DCA 003766)

Geronimo (mounted, left) and Naiche (mounted, right) with Perico (holding infant) and Tsisnah (far right). (Photograph by Camillus S. Fly, March 26, 1886, courtesy of National Archives and Records Administration, NAID 533085)

James "Santiago" McKinn (front) had been captured at age eleven by Geronimo's band. When told that he was to be returned to his parents, he answered in Apache "that he didn't want to go back—he wanted to always stay with the Indians," and "acted like a young wild animal in a trap." (Photograph by Camillus S. Fly, March 26, 1886, courtesy of Library of Congress, Prints and Photographs Division, 2006682475)

Geronimo (foreground, third from left) and General George Crook (foreground, second from right) in conference at Cañon de los Embudos, Sonora, Mexico, March 25, 1886. John Gregory Bourke is immediately to the left of Crook; Charles Roberts, age thirteen, is to the right; and Crook's mule, Apache, is in the background, center. (Photograph by Camillus S. Fly, courtesy of Palace of the Governors Photo Archives, NMHM/DCA 002116)

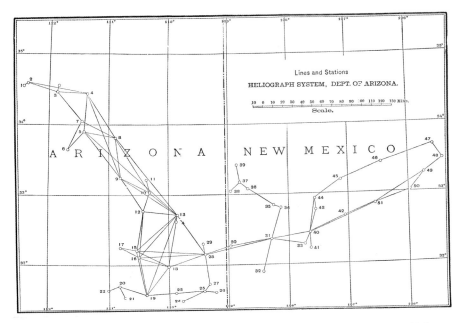

U.S. Army heliograph network, 1891. (After Nelson A. Miles, *Personal Recollections*, 1896)

Chiricahua Apache prisoners en route to Fort Marion, Florida, at Nueces River, Texas, September 10, 1886. Geronimo is in the front row, third from the right; Naiche, center. Lozen, a clairvoyant warrior and the younger sister of Victorio, is in the back row, third from the right. (Photograph by A. J. McDonald, courtesy of National Archives and Records Administration, NAID 530797)

FIG. 1.

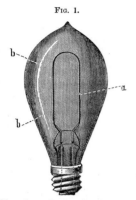

Glow lamp, having the glass bulb blackened by deposit of carbon, showing the molecular scattering which has taken place from the point *a* on the filament, and the shadow or line of no deposit produced at *b*.

John Ambrose Fleming, who coined the term "electronics," observed in 1890 that the filament of an electric lamp radiates light uniformly in all directions, while some as yet unidentified charged particles, soon to be known as electrons, are emitted asymmetrically within the evacuated bulb. (John Ambrose Fleming, *Problems in the Physics of an Electric Lamp*, February 14, 1890)

FIG. 4.

The "Edison effect." Seeking to understand this asymmetric conductivity within the vacuum, Fleming inserted a small metal plate into an incandescent lamp, leading to his invention of the vacuum tube, or, as he termed it, the thermionic "valve." (John Ambrose Fleming, 1890)

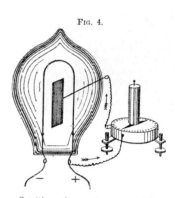

Sensitive galvanometer connected between the middle plate and positive electrode of a glow lamp, showing current flowing through it when the lamp is in action ("Edison effect").

Lee de Forest's Audion, which "magnifies and translates into sensation electric energies whose very existence as well as form and frequency, would but for it remain utterly unknown," included a third electrode made out of platinum wire bent into the shape of a fireplace gridiron, allowing a small control current applied to the grid to regulate the flow of electrons from cathode to anode within the tube. A long-plate version, ca. 1914, is shown here. (Courtesy of John Jenkins, SPARK Museum of Electrical Invention)

Freeman Dyson on the upper River Wye, age six. "Imagine a child playing in a woodland stream, poking a stick into an eddy in the flowing current, thereby disrupting it," explained the biologist Carl Woese. "But the eddy quickly reforms. The child disperses it again. Again it reforms, and the fascinating game goes on. There you have it! Organisms are resilient patterns in a turbulent flow." (Courtesy of Alice Dyson)

Sir Phillip Roberts's Erolunar Collision, an unfinished account of a voyage to the moon to observe a collision with the asteroid Eros, written by Freeman Dyson, age eight in 1932. (Freeman Dyson papers, American Philosophical Society)

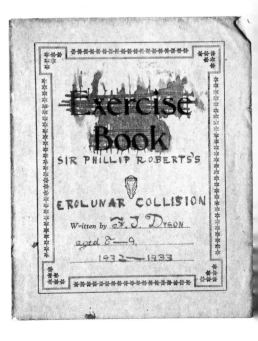

Exercise Book

SIR PHILLIP ROBERTS'S

EROLUNAR COLLISION

Written by F. J. Dyson
aged 8—9.
1932—1933

Verena Haefeli (neé Huber) behind the wheel of her 1940 Dodge, with Freeman Dyson and Fritz Rohrlich, at the New Jersey shore, May 8, 1949. (Photograph by Cécile Morette)

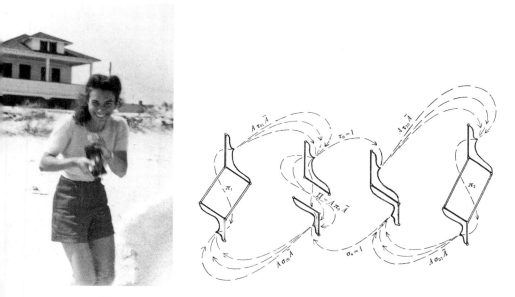

LEFT Verena Haefeli, May 8, 1949. (Photograph by Cécile Morette)

RIGHT Diagram from Verena Haefeli's 1947 dissertation on the classification of finite groups. "Looking at my Ph.D. thesis makes me wince," she would write fifty years later. "It was a preposterous piece of mathematical speculation flailing wildly in the idiom of a cumbersome symbolism toward the goal of a structural analysis in finite group theory that was right on."

Leo Szilard and Gertrud Weiss Szilard at Memorial Hospital, New York, October 1960, with the tape recorder into which Leo dictated *The Voice of the Dolphins*. (Photograph by John Loengard for *Life* magazine, courtesy of John Loengard, Trude Szilard, and Special Collections, UCSD Libraries)

Theodore "Ted" B. Taylor and the central library and cafeteria at General Atomics, La Jolla, California, November 1999. The library was the same diameter as the four-thousand-ton interplanetary vehicle proposed to ARPA in 1958.

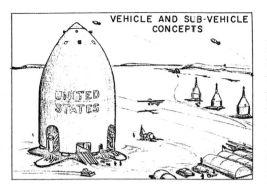

Multiple Orion Mars exploration mission: two four-thousand-ton Orion ships remain in Mars orbit in the background while the payload module of a third ship, separated from its propulsion module, has been landed to serve as a surface base. Three landing/return vehicles are on the right; note the bulldozer at the lower left for scale. (Sketch by Walter Mooney, June 25, 1963, courtesy of General Atomics)

Deep Space Force. Multiple independently targeted re-entry vehicles are launched by a four-thousand-ton Orion battleship that has deorbited from its station in deep space and entered a hyperbolic earth encounter trajectory to execute a retaliatory strike. (Artist unknown, from GA-C-962, *Potential Military Applications*, March 1, 1965. Courtesy of U.S. Air Force Special Weapons Center, Kirtland History Office)

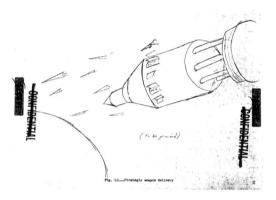

Freeman Dyson (right) watching preparations for tethered test of a one-meter-diameter high-explosive-propelled flying model at General Atomic's Point Loma test site, summer 1959. Clockwise from top: Ed Day, Walt England, Brian Dunne, Perry Ritter, Jim Morris, Michael Feeney, W. B. McKinney, Michael Ames. (Photograph by Jaromir Astl)

LEFT Margot Einstein, Elsa Einstein, and Helen Dukas in Old Lyme, Connecticut, 1935. (Photographer unknown, courtesy of Shelby White and Leon Levy Archives Center, Institute for Advanced Study, EB 160)

RIGHT Norman Clyde, age eighty-two, in camp below the Evolution Range in the Sierra Nevada, after a career that included 118 first ascents. (Photograph by Verena Huber-Dyson, 1967)

John Brower, David Brower, and George Dyson, with a twenty-eight-foot three-hatch baidarka at the Starboard Light Lodge, Belcarra Park, British Columbia, 1978. (Photograph by Richard C. Elson)

Crew of the *D'Sonoqua*; George Dyson at the far left, 1971. The Sitka spruce masts were hauled out of the woods at Kewquodie Creek near Mahatta River in Quatsino Sound; the standing rigging was contributed surreptitiously by a BC Hydro maintenance crew; the topsail was a canvas hatch cover salvaged from the historic RCMP arctic patrol vessel *St. Roch*. (Photographer unknown)

Goat aboard the *D'Sonoqua*, Desclation Sound, British Columbia, 1971. Many homesteaders kept milk goats, but billy goats were scarce. One of our assignments, as a supplier of trade goods to the islands, was to take this billy on the rounds of Desolation Sound.

Belcarra Park shoreline, Burrard Inlet, British Columbia, with the tree house barely visible in the tall Douglas fir appearing against the skyline on the left. (Photograph by Ann Yow, 1980)

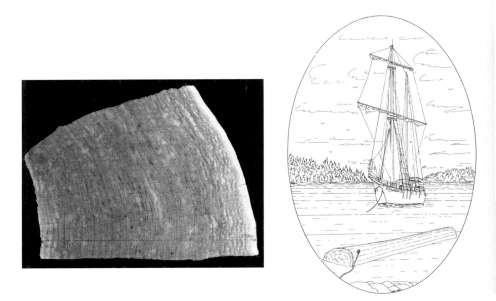

LEFT Cross section of red cedar shake bolt, showing growth rings going back to the year 1426.

RIGHT The *D'Sonoqua* at anchor off Belcarra Park, August 1972. The drift log that would become a tree house has just been towed ashore.

RIGHT Tree house, viewed from an adjacent tree, 1974.

BELOW View from the tree house, looking northwest across the entrance of Indian Arm toward Deep Cove, 1973.

Oliver Thomas, George Dyson, and Claudia Thomas in the doorway of the tree house, March 1975. (Photograph by Peter Thomas)

LEFT Tree house, viewed from Burrard Inlet. Freshly split red cedar boards, being used to finish paneling the interior, are visible at the base of the tree, ninety-five feet below. (Photograph by Ror. Orieux, June 1973)

BELOW Beth Yow, Ann Yow, and Laura Yow, August 1981.

BOTTOM "Trout" stove made by the Albion Stove Works, Victoria, British Columbia, and salvaged from a sunken vessel. (Photograph by Ann Yow) persuaded the San Diego city fathers to grant General Atomic 320 acres at the city limits in exchange for the riches the atomic age would bring. In the early spring of 1958, General Atomic began moving to its new headquarters at Torrey Pines.

The new laboratory was centered on a circular technical library with a courtyard in the middle and a cafeteria on the upper floor. The building, resting on steel columns around its periphery that could stand in for shock absorbers, was 135 feet in diameter: the exact dimensions of the four-thousand-ton Orion design. Taylor would point to a car or a delivery truck, the size of existing space vehicles, and say, "This is the one for looking through the keyhole." Then he would point to the library and say, "And this is the one for opening the door."[43]

For the first six months the project was kept so secret that its existence could not be revealed even to family or friends. "Things are happening which will involve me more and more, and may end up taking me away from Princeton for half my time or more," Freeman Dyson wrote to Verena Huber-Dyson on April 1, 1958. "I might even be away for several years. The call is just the one that I cannot resist. You know me too well not to guess what all this is about."[44]

In May 1958, General Atomic received permission to reveal the existence of the project for recruiting purposes, and on July 2, 1958, a one-page press release announced "the possible development of a new concept of propulsion employing controlled nuclear explosions . . . within the atmosphere and beyond."[45] I could now be told why my father kept disappearing out west. "When I drove George to school this morning I told him about the space-ship," our au pair Imme Jung reported in June 1958. "He was very excited, asked immediately which planet you will send it to, and whether there would be a little seat right next to you for him to come along."[46]

Over the July Fourth holiday in 1958, Dyson sent a five-page letter to Oppenheimer asking for a leave of absence to stay in La Jolla and join the project full-time. "Either through a misunderstanding of our intentions, or by an unusual and deliberate act of wisdom, the government has announced to the public that we are working on the design of a space-ship to be driven by atomic bombs," he wrote. "This scheme alone, of all the many space-ship schemes which are under development, can lead to a ship adequate to the real magnitude of the task of exploring the Solar System. We are fortunate in that the government has ordered us to go straight ahead for the long-range scientific objectives of interplanetary travel, and to disregard possible military uses of our propulsion system."

He added a statement of principles:

1. There are more things in heaven and earth than are dreamed of in our present-day science. And we shall only find out what they are if we go out and look for them.
2. It is in the long run essential to the growth of any new high civilization that small groups of men can escape from their governments and go and live as they please in the wilderness. Such isolation is no longer possible on this planet.
3. We have for the first time found a way to use the huge stockpiles of our bombs to better purpose than for murdering people.[47]

"I had an impression of what you were up to that seems to have been quite reliable," Oppenheimer replied, approving the request.[48] Dyson's appointment at General Atomic included a palatial Tudor-style house, draped in bougainvillea with a swimming pool and citrus orchard, above the surfing beach at Windansea, where a counterculture was emerging that would be immortalized by Tom Wolfe in *The Pump House Gang*. In September 1958, I entered first grade in a one-room schoolhouse overlooking La Jolla Cove. Oranges ripened in the winter sun, and dates dropped from the palm trees overhead. I believed the big round building behind the security gates at Torrey Pines was the launchpad for the spaceship and that one day, when it was finished, we would climb on board and go. In Ted Taylor, Freeman Dyson found the brother he never had. "Ted and I will fly together to Los Alamos this evening," he reported to his parents on July 1, 1958. "We travel like Paul and Barnabas. Golly, this life is good."

During World War II, Dyson had been kept on the ground. Project Orion was his chance to fly. He shifted his attention from physics to engineering, and because there was no existing brotherhood of nuclear-bomb-propelled spaceship engineers to defend the territory, he and Taylor had the field to themselves. "In the course of his deciding whether to stay with the project and resign from the institute, he said he had to make a choice between being a very good theoretical physicist or the best engineer ever," Taylor recalls.[49]

Dyson had an old brass telescope belonging to the era of Jules Verne. Soon after our arrival in La Jolla, he bought a new telescope and replaced our 1949 Ford with a 1957 turquoise-and-white Chevrolet Bel Air. The tail fins that were prominent on the Chevrolet were absent from Orion, which would have lofted itself so slowly through the atmosphere that tail fins would have been as ineffective as they were on the Chevrolet. After dinner at the Taylors', we sat outside as it got dark, taking turns looking at Mars, Saturn, and Jupiter while Taylor and Dyson discussed where they intended to go and when. The plan was to launch a four-thousand-ton ship on a shakedown cruise to Mars in 1965, taking off either from the existing atmospheric test site at Jackass Flats in Nevada or from a barge in the Pacific that could be washed down or sunk after the launch.

The first two hundred explosions, fired at half-second intervals, would lift the ship from sea level to 125,000 feet. Each kick adds about twenty miles per hour to the ship's velocity, an impulse equivalent to dropping the ship from a height of fifteen feet. Six hundred more explosions, gradually increasing in yield, would loft the ship into a three-hundred-mile orbit around the earth. "I used to have a lot of dreams about watching the flight, the vertical flight," remembers Taylor, who planned to be on board for the initial launch. "The first flight of that thing doing its full mission would be the most spectacular thing that humans had ever seen."[50]

Scientists and engineers from fourteen different countries joined the project, led by Taylor's conviction that national boundaries would dissolve in space. A crew of fifty to a hundred, divided evenly between air force officers and civilian scientists, would land on Mars for a year of surface exploration before returning to low earth orbit, where the ship could be refurbished or retired as an orbital base. Much of the mass that would be expended on the voyage was inert propellant that served, once the earth's atmosphere was left behind, to translate the energy released by the nuclear explosions into the momentum required to drive the ship. If a destination was chosen where propellant could be gathered for the return trip, a more ambitious mission could be attempted with the same takeoff mass.

The moons of Saturn promised both suitable propellant and a velocity match with the arriving ship, allowing a landing with minimal deceleration and without the losses to gravity required to land on Mars. "We knew very little about the satellites in those days," says Dyson. "Enceladus looked particularly good. It was known to have a density of .618, so it clearly had to be made of ice plus hydrocarbons, really light things which were what you need both for biology and for propellant."[51] In August 1958, General Atomic issued a twelve-page classified report, *Trips to Satellites of the Outer Planets*, outlining the size of ship, velocity increments, nuclear yields, and propellant budget that would be required for the trip.

The first-generation, twenty-kilometer-per-second ship designed to land on Mars could become a forty-kilometer-per-second ship that would land on Enceladus, replenish its expended propellant, and return to earth. "Saturn by 1970" became the motto of the group. "The Orion system, peculiarly well suited to take maximum advantage of the laws of celestial mechanics, makes possible round-trips to satellites of Jupiter in 2 years, or to satellites of Saturn in 3 years, with takeoff and landing on the ground at both ends," Dyson concluded. "Using the outer planets as hitching-posts, we can make round trips to their satellites with overall velocity increments which are spectacularly small. The probability that we can refuel with propellant on the satellites makes such trips hardly more formidable than voyages to Mars."[52]

"The mission was the grand tour of the solar system," remembers

Harris Mayer, who joined the project from Los Alamos. "But we also thought of it as a real commercial enterprise, because you could bring back things from space to the earth. At that time, 1958, we were not worried about taking off from the ground. We knew how to make nuclear explosions in the atmosphere, and the characteristics are different than in space and we could take advantage of it."[53]

Did Harris Mayer intend to go along? "I was chicken," he admits. "I knew enough about the space business, and this was dangerous. You had to be crazy to go." Did Dyson intend to go? "Oh, yes, he wanted to go," says Mayer. "Enceladus was the one he wanted to go to then. What was remarkable is that he would come and talk to me and it was as if everything were all done."[54]

On October 31, 1958, after a flurry of last-minute explosions on both sides, the United States and the U.S.S.R. entered into a moratorium on nuclear testing that lasted until September 1, 1961, when the Soviets broke the cease-fire with the first of forty-five explosions in sixty-five days. Fourteen of these yielded above a megaton, with one yielding sixty-three megatons, an explosion so large that it blew a hole in the atmosphere, radiating much of its energy into space. The moratorium had worked in Orion's favor: with testing on hold, talent at the weapons labs was idled and General Atomic could pick up the slack. To some, Project Orion was an interplanetary spaceship; to others, it was the ultimate in weapons-effects tests: keeping some of the best physicists working on improved bomb designs and computer codes while the moratorium was in effect. "We should think of Orion as paving its own way through the nuclear explosion test ban, rather than simply waiting for it to disappear," Taylor noted at the time.[55]

American politics, not international treaties, brought Orion to a halt. The project was orphaned almost from the start. The military was unable to adopt a project aimed at exploring the solar system, while NASA was unwilling to adopt a project driven by bombs. For a brief moment in history ARPA was responsible for both. Dyson and Taylor envisioned a NASA-USAF collaboration, the way James Cook had explored the Pacific in Royal Navy vessels, or the U.S. Navy provides Antarctic logistics while the National Science Foundation supplies the scientific mission and personnel. They underestimated the bitter rivalries over space. It was NASA's refusal to endorse an interplanetary mission in ships that would belong to the U.S. Air Force that kept the project on the ground.

In September 1959, Dyson resigned from General Atomic, and we returned to the lowlands of central New Jersey, with the peaked tail fins of our Chevrolet the only tangible reminder, besides a handful of charge-canister fragments that my father kept in a desk drawer at his office, of the dreams that had been left behind.

The Air Force Special Weapons Center (AFSWC) in Albuquerque kept Orion on life support, shielding the project from those who sided with NASA in trying to shut it down. "Orion was a way to put a burr under the air force saddle blanket," explains Brian Dunne. "When you would go out privately with people in the air force, and talk about what's Orion for," adds Taylor, "it was to explore space, no question about that." To fend off NASA and justify allocation of air force funds, AFSWC required a military mission. They named it Deep Space Force.

By 1960, the global nuclear stockpile had reached thirty million kilotons: ten thousand times the total firepower expended in World War II. The policy known as mutual assured destruction, or MAD, depended upon the survival of sufficient retaliatory forces to deter a first-strike attack. Deep Space Force would have shifted a critical fraction of the U.S. strategic nuclear forces to a fleet of Orion ships stationed in orbits beyond the moon, allowing more time to consider or reconsider a retaliatory strike.

"On the order of 20 space ships would be deployed on a long-term basis," a declassified summary explains. "At these altitudes, an enemy attack would require a day or more from launch to engagement. Assuming an enemy would find it necessary to attempt destruction of this force simultaneously with an attack on planetary targets, initiation of an attack against the deep space force would provide the United States with a relatively long early warning of an impending attack against its planetary forces. Because of its remote station, the force would require on the order of 10 hours to carry out a strike, thereby providing a valid argument that such a force is useful as a retaliatory force only. This also provides insurance against an accidental attack which could not be recalled." Crews of twenty to thirty, enjoying "an Earth-like shirt sleeve environment with artificial gravity," would be "deployed alternately, similar to the Blue and Gold team concept used for the Polaris submarines" in orbits about 250,000 miles from earth.[56] General Atomic would build the ships, the same way that the Electric Boat Division of General Dynamics built nuclear submarines.

Orion, cloaked in secrecy, became a pawn in a rivalry that went back centuries: whether wars should be decided by admirals commanding ships at sea or by generals commanding armies on the ground. The army, which had enlisted Wernher von Braun, won. Instead of Deep Space Force and voyages to Mars and Saturn, we planted the flag on the moon using von Braun's rockets and then ran aground in Vietnam. Sixty years after Sputnik, the flagship product of General Atomic is not Orion, the manned interplanetary spaceship; it is Predator, the unmanned drone.

President Kennedy bristled at General Thomas Power's statement, as commander of the Strategic Air Command, that "whoever controls Orion will control the world." AFSWC was reprimanded for allowing General Atomic to venture beyond feasibility studies and never received authorization to proceed with nuclear tests. After the limited nuclear test ban treaty was signed in 1963, Ted Taylor kept up the hope of launching an interplanetary mission as a treaty-compliant joint venture with the U.S.S.R. Niels Bohr lent his support. But without support from NASA, this proposal stalled.

By February 1965 the Orion group at General Atomic was down to nine, and on June 30, 1965, Major John O. Berga of the Air Force Weapons Laboratory issued a terse, one-page Plan Change, concluding, "This Project is hereby terminated."[57] Orion, caught between the hopes of the 1950s and the fears of the 1960s, was at an end. The peaceful mission was publicly abandoned, while military programs that originated with Project Orion—directed-energy nuclear weapons, electromagnetic-pulse weapons and countermeasures, and anti-ablation defenses against anti-satellite weapons—have secretly continued to this day.

Freeman Dyson returned to La Jolla for a sabbatical year in 1964–1965, but had nothing further to do with Project Orion except to write an obituary, published in *Science* under the title "Death of a Project," with a one-sentence abstract: "Research is stopped on a system of space propulsion which broke all the rules of the political game." Only now could he write publicly about a project that should have been in the open all along. Orion's opponents resented the attention and made sure that any chance of restoring funding was withdrawn. One of these critics "clearly wishes ORION not only to stay dead, but quiet also," a memorandum from the project's advocates at AFSWC complained.[58]

"The men who began the project in 1958," Dyson reported, "aimed to create a propulsion system commensurate with the real size of the task of exploring the solar system, at a cost which would be politically acceptable, and they believe they have demonstrated the way to do it."[59] But once NASA chose the path of chemical rockets, "there was no more brave talk of manned expeditions to Mars by 1965, and of sampling the rings of Saturn by 1970. What would have happened to us if the government had given full support to us in 1959, as it did to a similar bunch of amateurs in Los Alamos in 1943? Would we have achieved by now a cheap and rapid transportation system extending all over the Solar System? Or are we lucky to have our dreams intact?"[60]

By 1964, the La Jolla I had known as a six-year-old was gone. Instead of a one-room schoolhouse, I faced a large, anonymous junior high school and made few friends. I spent most of my time alone: building balsa wood airplanes, flying kites, and exploring the canyons above Scripps Beach. I remember skateboarding to school in the morning, but I cannot remember ever doing homework when I got back. One of my kites spent a continuous seventy-two hours aloft.

At regular intervals Trude Szilard signed me out of school early and took me to lunch at the Mediterranean Room of La Valencia Hotel, overlooking La Jolla Cove. She would order a pink martini and a shrimp salad while I nibbled at my food, served with memorable French bread. Trude, who was a practicing physician but had begun her studies as a physicist, sometimes smoked a pinch of dark tobacco out of a small silver pipe. "I can imagine that she found you to be in some ways a substitute for Leo," answered my father, when I asked him how these lunches had been arranged. "My guess is that Trude invited you because she enjoyed the company of a bright eleven-year-old with rebellious ideas."

Trude and Leo, who had met in 1929, maintained an unorthodox relationship that was formalized by marriage only in 1951. In February 1964, after Leo's recovery from cancer, they moved to La Jolla, taking up residence at La Valencia Hotel. Leo was awarded a fifteen-thousand-dollar salary and a small research budget by the Salk Institute at Torrey Pines, where he occupied an office in the cliff-top headquarters that had been designed by Louis Kahn. The Salk Institute was a refuge for biologists, neuroscientists, and physicists, conceived by Szilard as an apology for his role in the invention of nuclear weapons, over which he held such regrets. The Hotel del Charro at nearby La Jolla Shores, once favored by patrons of the Del Mar racetrack but now favored by consultants to General Atomic, had a stable that was no longer needed for racehorses, and this was converted to an apartment where Leo and Trude unpacked their suitcases and settled down for the first time in their lives. On May 30, 1964, Leo died in his sleep of heart failure, and Trude was left on her own.

She gave me a copy of *The Voice of the Dolphins*, a manifesto on the nuclear predicament, disguised as science fiction, that was dictated during Leo's hospitalization in New York, transcribed and edited by Trude, and published in 1961. Frustrated at the ineffectiveness of rational arguments against the insanity of nuclear strategy, Leo told a story, beginning in 1960 and ending in 1985, in which a delegation of dolphins intervene to reverse the nuclear arms race where human beings had failed.

Leo sent a Russian translation to the Soviet premier, Nikita Khrushchev, who agreed to meet privately in October 1960 in New York. The meeting, scheduled for fifteen minutes, extended into a two-hour discussion of how to prevent, or at least de-escalate, a nuclear war, including Szilard's suggestion to establish an open telephone line between the Kremlin in Moscow and the White House in Washington, D.C. "When I last enquired how long it might take for the President of the United States and the Chairman of the Council of Ministers of the Soviet Union to contact each other over the telephone in case [of] an accidental or unauthorized attack," Szilard had noted to Khrushchev before the meeting, "I was told that this might take several hours."[61] The Moscow-Washington hotline was established in 1963.

Szilard presented his guest with a Schick Injector razor and a six-month supply of blades, promising to send more as long as the United States and the U.S.S.R. avoided war. Khrushchev accepted the razor, adding that "if there is a war he will stop shaving, and he thinks that most other people will stop shaving also."[62] He offered a case of vodka in return, but Szilard, suffering from bladder cancer, asked for mineral water instead. Two cases, along with caviar and three smoked fish, were delivered to his room at Sloan Kettering, where he was supervising his own radiation treatment at the time. Szilard kept his promise, replenishing the supply of razor blades until, in the aftermath of the Cuban missile crisis, the channels of communication closed.

Szilard then took his campaign to Washington, staying at the Dupont Circle Hotel. "He lobbied for nuclear disarmament by sitting in a stuffed chair in the lobby of the Dupont Hotel just off Dupont Circle," remembers Ted Taylor, who was working for the Defense Nuclear Agency at the Pentagon at the time. "He had one of these racks you can turn, full of *The Voice of the Dolphins*, and he'd grab people by the coat sleeves as they went by, and talk to them about disarmament. He gave them an autographed copy of *The Voice of the Dolphins*. He'd spend the whole day, except for lunch, sitting in the lobby, lobbying. He died within a year or two; that was sort of his last cry."[63]

In Szilard's tale, the dolphins, after it is "gradually revealed that their intelligence far surpassed that of man," begin collaborating with a group of scientists, guiding them toward solutions, including nuclear disarmament and universal birth control, that save humanity from itself.[64] He claimed the story was about the stupidity of human beings, not the intelligence of dolphins. I saw it the other way around.

In September 1965 we returned to Princeton, where I built a thirteen-foot canvas-covered, wood-framed kayak, imagining arctic adventures but with nowhere to paddle except short distances up and down the abandoned Delaware and Raritan Canal. Now in high school, and bored, I escaped from school at least one day a week and, in winter, retreated into the subterranean depths of Princeton University's Firestone Library until school got out. There were eight miles of stacks. Firestone was closed to outsiders, but if you melted in with the morning influx of students, you could get swept inside. In one of the basements was a small reading room where some benefactor's collection of books on exploration and adventure had been kept intact, along with an old armchair, where I read the classic arctic voyages and other tales of the far north.

My mother, who had moved to Los Angeles in 1964 for a research position in "mathematical logic, recursive function theory and the theory of automata" with the Ground Systems Division of Hughes Aircraft, began having second thoughts about working for a defense contractor and moved north to join the logicians at Berkeley, where a separate department of computer science did not yet exist. Half the math department were mountaineers. She joined the Sierra Club and in the summer of 1967 signed us up for a mountaineering base camp at twelve thousand feet in the High Sierra, where I saw real wilderness for the first time. The next summer, at a base camp in the Evolution Basin, I was hired on as a scullion, the lowest position in the commissary crew, under the head cook, Barbara Brower, for twenty-five dollars a week.

Two legendary California mountaineers—David Brower, born in 1912, and Norman Clyde, born in 1885—were among the group in camp. They were the direct successors to John Muir. Brower had attempted the first ascent of Mount Waddington in British Columbia, from sea level at the head of Knight Inlet, in 1935, and made the first ascent of Shiprock in New Mexico in 1939. He helped train the U.S. Army's Tenth Mountain Division during World War II, earned a Bronze Star in the liberation of Italy, and transformed the Sierra Club into a modern environmental organization and a groundbreaking publisher while keeping the outings program true to its mountaineering roots. Norman Clyde, with 118 first ascents in the Sierra to his credit, still carried a crosscut saw, a four-pound ax, and a cast-iron skillet in his wood-framed pack. He was a noted sharpshooter with a pistol and could catch trout with his bare hands. The last time I passed him on the trail, he was eighty-three, I was fifteen, and we were both making our way over 12,800-foot New Army Pass.

I returned to a Princeton in upheaval over U.S. involvement in Vietnam. Local high school students were unwelcome on the university campus, but we knew how to sneak into most of the buildings and slipped through a fire exit into the back of Murray Dodge Theater to hear Muhammad Ali, recently stripped of his world heavyweight title over his defiance of the Selective Service Board, speak to the injustices of the war. Eighty years after the U.S. Cavalry, mounted on horses, had fought the last of the Indian Wars, our cavalry, now mounted on helicopter gunships, were pursuing guerrilla fighters in Vietnam.

I left high school without graduating in June 1969, headed west to Denver, and arrived in a snowstorm at Loveland Pass, intending to walk south along the crest of the Continental Divide to the headwaters of Vallecito

Creek in the San Juan Mountains north of Durango, where I had another summer job with a Sierra Club base camp lined up. Halfway through the trek, I descended and hitchhiked back to Boulder to watch the televised Apollo moon landing on July 20, 1969. Gone were the hopes that Mars and Saturn would be next. I knew enough about the fate of Project Orion to realize that the moon landing was not so much a beginning as an end.

In September, I enrolled in the University of California at San Diego, still operating informally enough that entrance to a sixteen-year-old without a high school diploma was allowed. After a single quarter I transferred to UC Berkeley, where I lived on a small sailboat in the city marina and spent more time sailing than attending class. Katarina, my older half sister, had joined the Quaker war resistance movement, moved to Vancouver, British Columbia, and invited me to her wedding, so in the spring of 1970, just as I turned seventeen, I headed north. Canada would soon put a fleet of Baffin Island kayaks on its two-dollar banknotes and a Johnstone Strait purse seiner on the five. It felt like home. I answered a newspaper advertisement for crew on a boat doing charter work on the B.C. coast, signed on, and didn't look back.

The vessel, unfinished, was named *D'Sonoqua*, and I was the only crew until it was launched. Jim Bates, the captain, was a welder and pipefitter by trade who had worked in coastal logging camps from the age of fourteen and knew Vancouver Island like the back of his hand. By sixteen he was setting chokers: dragging slings of heavy wire rope with a constricting loop at one end up the unstable freshly cut sidehills so that the marketable timber, waiting to roll over and crush you, can be yarded out from the surrounding debris. His father ran the Bates Esso service station in Parksville, where the logging roads to the north and west coasts of Vancouver Island branch off from the Island Highway, which at that time terminated at Kelsey Bay at the southern entrance to Johnstone Strait.

Bates was as puzzled by a seventeen-year-old who had never used a chain saw or started a diesel engine in his life as I was by the tasks I was assigned. Having been raised by a field theorist and a group theorist, I was in awe of those who wielded arc welders and cutting torches as effort-lessly as others use a knife and fork. Somehow I made myself useful, and with only one visit to the emergency room of Vancouver General Hospital the *D'Sonoqua* was launched.

Our first charter was a commission from the UBC neuroscience professor Paul Spong, stationed for the summer on Hanson Island, two hundred miles northwest of Vancouver, where he was establishing contact with a resident population of killer whales. Spong, a New Zealander, had been doing post-doctoral research at UCLA in behavioral neurophysiology when he received a joint invitation from the department of neuroscience at UBC and the Van-couver Aquarium, which were establishing a research program involving Skana, the first killer whale to survive in captivity for any length of time.

Spong had been working on vision in cats. Because almost nothing was known about vision in whales, cats were a good place to start. His proposal to study Skana's visual discrimination was accepted, and he got the job. He soon discovered that Skana was testing his perception as a scientist as much as he was testing her perception as a whale. She began giving wrong answers and seemed to be doing so just to see what he would do next. Skana was bored. To a killer whale, the Vancouver Aquarium was a sensory deprivation tank. Spong began attracting too much attention in his efforts to make life more interesting for Skana, and the administrators at the aquarium were relieved when he decided he would go and study her relatives in the wild instead.

After consulting with James Sewid, the Kwakwaka'wakw leader based in Alert Bay on Cormorant Island, across from the mouth of the Nimpkish River on northern Vancouver Island, Spong, with his wife, Linda, infant son, Yashi, an Afghan dog, and a calico cat, set up camp in a small bay near Burnt Point on Hanson Island, where the southern end of Queen Charlotte Strait converges with the northern end of Johnstone Strait in Blackfish Sound. Skana had responded to live musical perfor-mances at the aquarium, and Dr. Spong decided to conduct an experiment by bringing live music to Blackfish Sound. That's where the *D'Sonoqua* came in, along with a band named Fireweed, a fixture of the Vancouver music scene at the time. They loaded their equipment on board, and we headed north. By August 18 we were surrounded by whales in Johnstone Strait.

Spong's research station, now OrcaLab, had been selected for its exposure to his subjects, not shelter from the winds and six-knot tides. It was a precarious place to anchor overnight. I slept on deck next to the anchor chain to raise the alarm if the anchor started to drag and was woken up by the pods of whales that came by during the night. I looked over the unfinished bulwarks into the phosphorescent water and they looked back. As the whales circled the boat, I could hear their underwater vocalizations directly through the still night air. Later, sleeping in a kayak as whales passed by underneath, I could feel a penetrating series of clicks as they scanned the length of the kayak, the way we humans now scan our own offspring, in the womb, with ultrasound.

After Spong established his base of operations on Hanson Island, a stream of researchers followed, believing that if they recorded enough whale communications, they could decipher the language of the whales. For fifty years the quest has failed. Whales both perceive their surroundings and communicate using sound, which behaves differently in an incompressible medium like water than in a compressible medium like air. If humans could communicate directly, brain to brain, using light, would we have developed languages based on a limited vocabulary of sounds? Human language, either evolved from or coevolved with sequences of discrete gestures, is optimized to withstand poor transmission over a noisy, low-bandwidth channel and might emerge quite differently among minds not subject to these constraints. Whales are no doubt communicating, but not necessarily by mapping their intelligence to sequences of discrete symbols the way we use language to convey our thoughts. When we play music, the whales might be thinking, "Finally, they are showing signs of trying to communicate like us!"

The difference between analog and digital computing parallels the question of whether a linear, symbolically coded language is a necessary indicator of conscious intelligence or not. In a digital computer, higher-dimensional inputs are reduced to one-dimensional strings of code that are stored, processed, and then translated back into higher-dimensional outputs, with a hierarchy of languages mediating the intervening steps. Large numbers of logical operations are transformed into waste heat along the way. Among analog computers, information can be stored, processed, and communicated directly as higher-dimensional maps.

At Hanson Island, Leo Szilard's vision had come to life. Killer whales (*Orcinus orca*) are the largest of the dolphins, evolved from land mammals who returned to the sea more than a hundred million years ago. They roam the entire planet as separate populations belonging to a single species, forming complex, persistent matrilineal social structures, with young males mentored by their post-reproductive grandmothers, whose life spans are known to reach a hundred years or more. Breathing, sleep, and other physiological functions are synchronized across the members of a pod. Their communication may be closer to telepathy than to language as we know it, and it could even be that orca mind and consciousness is a parallel, distributed property belonging to the pod collectively as much as to any individual whale. In Szilard's story the dolphins make contact with a group of scientists, connected to the highest levels of government, who put their advice into effect. In 1970 at Hanson Island, they made contact with Dr. Spong.

When Spong showed up for work in Vancouver, his new boss, Patrick McGeer, reached into a laboratory freezer and brought out the brain of Moby Doll, Skana's unfortunate predecessor. The brain was as complex and convoluted as a human's but weighed 6.45 kilograms—more than four times as much.[65] What kind of mind might arise, over millions of years, among a society of large, highly developed brains immersed in a high-bandwidth telecommunications medium? Chattering, hand-waving primates are trying to find out.

If the Szilards were writing *The Voice of the Dolphins* for the fourth-epoch future of today, the cetaceans might establish contact not with a group of human scientists, still searching for a coded, one-dimensional language, but with an intelligence able to communicate directly, unimpeded by language, with a mind held in common by the whales.

5

TREE HOUSE

On the coast of British Columbia, sixteen thousand miles of shoreline are insinuated into the six hundred miles that separate the Strait of Juan de Fuca in the south from Dixon Entrance in the north. To visit all forty thousand islands would take more than a hundred years, if you slept on a different island every night. A maze of inlets is swept by sixteen-foot tides, and at high water you can pilot a deep-draft vessel so close to the overhanging forest that you risk entangling the shrouds. Even when following a mid-channel course, you meet up with floating logs, as if the forest were diffusing into the ocean the way that water evaporates, one molecule at a time, into the air.

Before the advent of commercial logging, drifting logs were rare. The only way for a log to leave the rain forest was as a boat. A few selected cedar trees were carved into canoes, split into planks for longhouses, preserved intact as house posts, or stripped periodically of their fibrous bark. During the first century of European settlement, the market for timber

View *of the* Entrance *of* NOOTKA Sound, *when the N point of the Entrance bore E. dist. 4 Miles.*

was limited to a small number of Douglas firs, taken whole for spars. Suddenly this changed. The first sawmills were established at Victoria in 1848, Port Alberni in 1861, and Moodyville (now North Vancouver) in 1862. By 1912 the British Columbia coast supported 250 sawmills and its first three pulp mills: at Powell River, Port Mellon, and Ocean Falls. The mills evinced an insatiable appetite for logs. Forests that had grown undisturbed for millennia were felled by loggers who swarmed through the inlets like ants.

Rainfall on the British Columbia coast is measured in hundreds of inches per year and intervals between fires in centuries. Until the loggers showed up, the rain forest was constrained only by gravity, its canopy reaching the height to which water can be lifted by osmotic pressure and there it stopped. Nowhere else on earth was it quite so easy to fell enormous firs and cedars within immediate reach of tidewater, where they were rafted up into booms and towed off to the mills that had metastasized to the far reaches of the coast. Making one or two knots through channels with six-knot tides, small coastal tugboats with names like *Fearless* and *Joe Drinkwater* cultivated an uncanny rapport with the local currents, capturing gravity assist from the moon through the diurnal tides.

The harvesting of the British Columbia rain forest began with opportunistic hand loggers who picked off the fir and cedar closest to the shoreline, followed by small-time operators who made off with a wider selection of timber as they shifted their floating bunkhouses, cookhouses, and donkey engines from one inlet to the next. Improbably long boom sticks, brow logs, and stifflegs defined the layout of the floating camps, their wet, slippery surfaces perforated by the spiked soles of loggers' caulk boots, linking everything together and connecting it to shore. Donkey engines were large, wood-fired steam engines, coupled to a winch whose drum held a long steel cable for hauling logs out of the woods. They were either operated beneath enormous A-frames floating on huge log rafts or landed ashore on skids to haul themselves through the forest, like some monstrous fire-breathing inchworm, one length of cable at a time.

As easy pickings grew scarce, larger companies established fixed camps and built railways reaching far inland. Diesel superseded steam, logging roads were carved into the landscape, and, where roads could not quite reach, spar trees were commandeered by high riggers who limbed the trunks on the way up to install the guylines, snatch blocks, mainlines, and haulbacks that cleared entire sidehills, leaving characteristic radiating scars resembling crop circles in a field of grain but on a vastly larger scale. There was no escape. But when the logs made it to the water, some found a way out.

Drift logs range across the spectrum of commercially exploited species—red cedar, yellow cedar, Douglas fir, hemlock, and Sitka spruce—and fall into two types: those that float horizontally, lying down, and those that float vertically, standing up. Those that float vertically, known as deadheads, are more hazardous to boats. A small deadhead one foot in diameter and thirty feet long, floating with only a few inches above the surface, will weigh almost a ton. In a seaway, it will pick up a vertical oscillation in resonance with but on a period much longer than that of the background waves. If a vessel coming down off a wave meets a deadhead on its way up, the boat may sink in minutes with a large hole punched through the underbelly of its hull.

Deadheads are a by-product of the logging industry, especially the harvest of second-growth timber for pulp. Trees harvested at maturity will float horizontally for years, but a young hemlock, heavy with sap, will start off with barely enough buoyancy to float. A log that sinks at one end may be swept away from a log boom by a passing current or lost while under tow. In times past there were more wooden vessels, more logs being stored and towed around in log booms, and many more sinkings caused by deadheads, especially at night. It was as if the forests had found a way to take revenge on those who had cut them down.

Working vessels have no choice but to run at night. Rocks are hazardous, but their positions are marked on the charts. Deadheads are distributed at random on a minefield with no map. If you are running at night on a wooden-hulled vessel, you wish you were on a steel-hulled vessel and breathe a sigh of relief at first light.

High-floating debris is less dangerous and tends to collect along the lines of demarcation between opposing surface currents, with small pieces of driftwood signaling the proximity of larger ones. Certain high-floating logs were valuable enough to be sought after, rather than avoided, and a species of freelance scavenger, commensal with the forest industry, developed to meet the demand. Under Canadian civil law the log salvor, or beachcomber, "is a finder of lost property and would be deemed to be the owner of such logs except as against persons who can show a better title."[1] The means by which the logging companies showed a better title than the beachcombers was the registered timber mark: a unique numbered symbol, similar to a livestock brand, registered with the provincial government and cast into the face of a heavy steel hammer with which the exposed end grain of a freshly cut log was given a vigorous blow. The impact displaces the longitudinal fibers, leaving an imprint that cannot be erased. Under the Criminal Code of Canada, up to five years' imprisonment could be imposed on anyone who "removes, alters, obliterates or defaces a mark or number on . . . any timber, mast, spar, shingle bolt, sawlog or lumber of any description . . . that is found adrift [or] cast ashore."[2]

To the lasting dismay of the beachcombers, the logging industry, with the collusion of the sawmills and the assistance of the Ministry of Forests, maintained both this efficient means of enforcing title and a system for scaling and buying back the salvaged timber at a price just high enough to keep the log salvors in business, but below market cost. It was illegal to sell salvaged logs anywhere else. After severe storms the beachcombers, who divided the territory among themselves, scoured the coast. But the timber companies stacked the deck: if a log boom broke up, they could declare the area closed to salvaging for as long as necessary to recover the spilled logs themselves before giving the beachcombers a chance.

Canada was a confederation of dichotomies, not only in its bilingualism but also in its embrace of central authority, from a monarchy to a national police force to a single broadcasting network, and extreme individualism

at the same time. The Hudson's Bay Company monopoly had depended
on the independent voyageurs, while the voyageurs, for all their apparent
independence, depended on the Hudson's Bay Company to market their
furs. The pattern had been set. You could buy alcohol only from the British
Columbia Liquor Control Board and you could sell beachcombed timber
only through a receiving station licensed by the Ministry of Forests, but in
between, if you were not distilling your own alcohol or pilfering from the
booming grounds, you were left alone. If the Mounties pulled you over on
one of their rare visits to the smaller islands, you were in serious trouble
only if they dipped a white handkerchief in your gas tank and found you
were burning purple gas, which was untaxed and only for use in boats.

In late August 1972, I was at the helm of the *D'Sonoqua* motoring across
Georgia Strait, somewhere between Lasqueti Island and Point Atkinson,
when we crossed paths with a large high-floating log. We had been running south all night, the lights of Vancouver still invisible below the horizon, lending the sky in the southeast a pale glow now blending into the
first light of dawn. The *D'Sonoqua* slid across a glassy sea, leaving a phosphorescent wake. As the twilight differentiated into a horizon, something
appeared off the port bow: a thirty-foot length of old-growth cedar, three
feet in diameter, its paper-smooth surface denuded of all bark and weathered to a pale blond. I pulled back on the throttle, took the engine out of
gear, and coasted up to the log as the captain climbed up the aft compan-
ionway and appeared on deck.

He took the wheel and threw the engine into reverse. I ran astern to get
a dog line and a small sledgehammer, grabbing our twenty-foot pike pole
from its place standing upright in the mizzen stays. We pulled the log in
alongside the *D'Sonoqua*'s waist, and I drove in a ring dog: a forged steel
spike with a large eye and a rectangular head on one end. Shaped like a
short, thick Tlingit dagger, it splits its way easily into the grain, but if you
drive it all the way in, past the shoulders of the spike, it won't let go unless
you have the right tool to extract it by twisting and pulling at the same

time. The pitch increases with every blow, signaling, sometimes for miles across the water at the first light of dawn, that someone has captured a log. With our prize in tow behind us, we continued on our way down the strait.

We entered Vancouver harbor at mid-morning with a flood tide, passing through First Narrows and under the Lions Gate Bridge, then another ten miles due east, squeezing through Second Narrows and under two more bridges as the tide began to turn. We dropped anchor in front of the Starboard Light Lodge: a 1920s-era homestead at the mouth of Indian Arm, north of Admiralty Point and now in Belcarra Regional Park. This would be a good place to leave the log. Taking beach wood or drift logs for personal use, though illegal, was part of life on the B.C. coast. Both provincial authorities and the local beachcomber, the only person with any reason to complain to them, tended to look the other way.

The city of Vancouver occupies a peninsula bounded by the Fraser River to the south and Burrard Inlet to the north. Burrard Inlet, along with its northern extension, Indian Arm, is the first of a series of fjords, carved by ice, that penetrate deep into British Columbia's still-glaciated Coast Mountains. At the head of Knight Inlet, only two hundred miles and five fjord systems north of Vancouver, glaciers, although rapidly retreating, still descend to within a thousand feet of sea level today. During the glacial advance the mainland inlets were excavated to depths of up to two thousand feet but were left much shallower at their entrances, where the glacial carving stopped, leaving a sill as the ice pulled back. In the depths of these underwater canyons live creatures separated from the rest of the ocean by shallows to which they cannot ascend, as isolated as populations of island tortoises on an archipelago turned upside down.

Indian Arm, known as North Arm until 1921, is more than seven hundred feet deep and has only two level sites anywhere on its shores. One is at the mouth of the Indian River at the northern terminus of the fjord, and the other is near the fjord's entrance, ten miles to the south, where a low, flat isthmus separates the narrow entrance to Indian Arm from Bedwell

Bay. This was the site of a large settlement known to its Coast Salish inhabitants, the Tsleil-Waututh, as Tum-Tumay-Whueton.

The archaeological evidence consists of two layers: one going back some one thousand to two thousand years, and the other much older, although how ancient remains unknown. Less than 1 percent of the site has been excavated, and much of the original beachfront midden has eroded away. The lowest layer of cultural material lies directly on top of thirteen-thousand-year-old glacial clay and till, suggesting that the earliest residents arrived soon after the departure of the Cordilleran ice sheet, under very different conditions from today.[3]

Thirteen thousand years ago the sea level was rising, uniformly, as the ice retreated, while the land was also rising, unevenly, with the removal of the overburden of ice. Most evidence of early coastal occupation is now either underwater or inland and overgrown. Known human occupation correlates with shellfish middens, but absence of shellfish middens does not imply absence of inhabitants. The first arrivals were once assumed to have been small bands of transient hunters who arrived on foot via the Bering Land Bridge from Asia, before spreading out inland along the interior fringes of the ice sheet and returning through ice-free corridors like the Columbia River and Fraser Canyon to the coast. Recent evidence suggests they were boatbuilding sea mammal hunters, following the "kelp highway" that extended from Asia along the southern shores of Beringia and then southward along the Northwest Coast.

Once salmon runs became established and the rain forest matured, a very different culture flourished on the Northwest Coast. The cultivation of resources that were fixed in place and predictable in time enabled large permanent settlements exhibiting complex social organization with advanced technology and arts. Systematic aquaculture and forestry enabled population densities, social institutions, ritualized warfare, and artistic display that was breathtaking to those who witnessed it firsthand. So many goods accumulated that a dedicated system, centered on the potlatch, described as "fighting with property," was instituted to formalize the redistribution of surplus wealth.[4]

The Fraser River estuary, teeming with cedar, salmon, shellfish, and waterfowl, supported a population, before a smallpox epidemic in 1782, that can only be guessed from the extent of the shellfish mounds that were left behind. The Great Fraser Midden at Marpole on the north side of the Fraser delta, surveyed by Charles Hill-Tout in 1892, was "upwards of 1,400 feet in length and 30 feet in breadth; and covers to an average depth of about 5, and to a maximum depth of over 15, feet an area exceeding 4½ acres in extent."[5] When Simon Fraser descended the river that bears his name in 1808, he described the inhabitants of the village of Musqueam on the north side of the estuary as occupying a "fort" of unbroken house-fronts "1,500 feet in length and 90 feet in breadth."[6]

Just north of the Fraser valley, at the foot of the Coast Mountains, the Belcarra site featured a broad canoe-landing beach in front, a clear view of any hostile raiding parties entering through Second Narrows from the west, and an alternate escape route northward into Indian Arm through Bedwell Bay. Cedar trees six feet and more in diameter, whose stumps are still in evidence, stood nearby. The tide flats were saturated with clams. Dense vegetation now conceals the barren landscape that had been exposed by the retreating ice. Yet perched at the high-tide line on the shoreline just north of the Starboard Light Lodge is a large erratic granite boulder, ten feet in diameter, a reminder left by a fast-moving lobe of the continental ice sheet that one day may return to pick it up.

The *D'Sonoqua*'s entrance through First and Second Narrows had been preceded, 180 years earlier, by the first European visitors to Burrard Inlet: a party of Englishmen led by George Vancouver and William Brough-ton, in the sloop *Discovery* and brig *Chatham*, followed by a party of Span-iards led by Dionisio Alcalá Galiano and Cayetano Valdés y Flores, in the schooners *Sutil* and *Mexicana*. The English were sixty-two weeks out from Falmouth and the Spaniards fourteen weeks out from Acapulco when the two expeditions arrived, just one week apart. The British entered and named Burrard Inlet on June 13, followed by the Spanish, who named it

Canal de Floridablanca, on June 21. Both parties were under instructions to search for a northwest passage linking the Northwest Coast to Hudson's Bay, a possibility that neither expedition placed much faith in, other than as a pretext to survey large areas of coastline that the earlier Russian, Spanish, and English voyages had overlooked.

This was Vancouver's second visit to the Northwest Coast. In January 1772, he had joined James Cook's second voyage to the Pacific as a midshipman aboard the *Resolution*, at age fourteen. On January 30, 1774, the *Resolution* reached 71°10' south, within 120 miles of the Antarctic continent, at which point, facing an impenetrable wall of ice, Cook noted that "we could not proceed one Inch further south."[7] Vancouver climbed out to the end of the bowsprit, just before the ship turned around, and claimed to have stood farther south than anyone else. The expedition returned home in July 1775, and in February 1776 Vancouver joined Cook's third voyage, this time as midshipman aboard the *Discovery* under Captain Charles Clerke.

On August 18, 1778, they reached 70°41' north, where they were once again stopped by ice. Vancouver, just twenty-one, had now reached the extremes of navigation in both hemispheres and would devote himself, in his own subsequent voyage, to completing the survey that Cook had begun on the Northwest Coast. When Cook was killed at Kealakekua Bay in Hawaii in 1779, it was Vancouver who led the party that recovered the body after his commander's death.

The Hawaiians had welcomed Cook as the embodiment of their divine Lono, but this illusion began to crumble as the English exercised the habits of eighteenth-century sailors and extended their stay to make repairs. "They regard us as a Set of beings infinitely their superiors," noted James King on February 4. "Should this respect wear away with familiarity, or by length of intercourse, their behavior may change." Pilfering increased, and on February 13, 1779, Vancouver, Thomas Edgar, and two other men took the *Discovery*'s small cutter in pursuit of a Hawaiian who had made off with a set of metal tongs from the *Discovery* and escaped by canoe to shore. The mood changed.

The English were unarmed and outnumbered by the Hawaiians, who attacked with stones and clubs. "All this time some of the Natives was Stoneing & beating the Midshipman and me, while the rest was Stealing the Oar's & furniture belonging to the Boat," says Edgar. "I not being able to swim had got upon a small rock up to my knees in water, when a man came up with a broken Oar, and most certainly would have knock'd me off the rock, into the water, if Mr. Vancouver, the Midshipman, had not at that Ins[tant] Step'd out of the Pinnace, between the Indian & me, & receiv'd the Blowe." One of the Hawaiian chiefs intervened to save Vancouver, and during the subsequent negotiations to recover Cook's body, King observed that the Hawaiians "seemed rejoiced to see Mr. Vancouver . . . who best understood them."[8]

Vancouver had spent four weeks with Cook at Friendly Cove in Nootka Sound in 1778, where the English commander and the Nootka chief Maquinna exchanged visits, shared meals, and treated each other as heads of state. Returning fourteen years later to extend Cook's survey into the Strait of Juan de Fuca, Puget Sound, and Georgia Strait, now collectively the Salish Sea, Vancouver found the landscape strangely underpopulated compared with the outer coast and the islands farther north. The number of empty, abandoned settlements "warranted an opinion that at no very remote period this country had been far more populous," he remarked.[9]

As they explored the Salish Sea, the British found open clearings "on the most pleasant and commanding eminences, protected by the forest on every side." The abandoned sites, overgrown with "nothing but the smaller shrubs and plants," pointed to a depopulation "occasioned by epidemic disease, or recent wars."[10] There were signs of smallpox among the survivors, suggesting an epidemic that originated in Mexico about a decade earlier and traveled north through the interior of the continent before making its way through river-based trade routes to the coast.

On June 11, 1792, Vancouver and Broughton brought the *Chatham* and the *Discovery* to anchor in Birch Bay, just south of the present-day boundary between Canada and the United States. They landed on the south side of the bay, near the site of "a very large Village now overgrown with a thick crop of Nettles & bushes," as the expedition botanist, Archibald Menzies, noted in his report.[11] The observatory was set up onshore to recalibrate the ship's chronometers, the blacksmiths were disembarked with their forge to effect repairs, while a contingent of sailors was assigned to collect wild greens and brew spruce beer as an antiscorbutic—the first brewing operation undertaken in what would become Washington State.

With the two ships secured, Vancouver and Peter Puget took the *Discovery*'s pinnace and launch, with a week's provisions, to survey the mainland inlets to the north. On the morning of the twelfth, they rounded Point Roberts, where they observed yet another "large deserted Village capable of containing at least 4 or 500 people,"[12] before traversing the delta of the Fraser River, whose shoals extended so far offshore that they were forced to head for the opposite side of Georgia Strait, where they landed (on Galiano Island) at one o'clock in the morning, after ten and a half hours at the oars. After four hours of rest, they rowed back across the strait, reaching the mainland at Point Grey at noon before continuing on across English Bay and entering Burrard Inlet through First Narrows, just north of the present Stanley Park. Tucked inside the headland was a midden eight feet deep and of such extent that when excavated, it provided enough clamshells to surface the entire nine-mile drive that was constructed around the perimeter of the park.

Proceeding up Burrard Inlet, they were met by "about fifty Indians, in their canoes, who conducted themselves with the greatest decorum and civility."[13] A century later, Andrew Paull, a living descendant, would recount to J. S. Matthews that "as your great explorer, Vancouver, progressed through the First Narrows, our people threw, in greeting before him, clouds of snow white eiderdown feathers which rose, wafted in the air aimlessly about, then fell, like flurries of snow, to the water's surface, and rested there like white rose petals scattered before a bride."[14]

Vancouver's party camped for the night on the south shore of Burrard Inlet, across from what is now Belcarra Regional Park. On their departure at four o'clock in the morning of the fourteenth, they passed "a small opening extending to the northward, with two little islets before it of little importance."[15] The opening was the entrance to Indian Arm.

Vancouver and Puget continued exploring the mainland coastline, surveying Howe Sound and Jervis Inlet before returning to Point Grey on June 21, where they met Galiano and Valdés, compared notes, and exchanged charts. The Spanish decided to take their own small boats and explore Indian Arm, where they noted several "meadows," including the Belcarra village site. The only inhabitants they observed were a few people at the mouth of the Indian River who scattered into the woods upon their approach.

After the departure of Galiano and Valdés, Burrard Inlet was neglected by Europeans for half a century while trading activity focused on the outer coast. The Spanish occupied California, while the Russians occupied Alaska, leaving the United States and England to compete for territory in between. To counter the encroachment of American traders, the Hudson's Bay Company established permanent outposts at Fort Langley on the Fraser River in 1827, at Victoria on southern Vancouver Island in 1846, and at Fort Rupert on northern Vancouver Island in 1849. The entire 275-mile-long island was leased to the company for seven shillings per year, while the mainland was largely ignored until gold was discovered in the British Columbia interior in 1858. Burrard Inlet was both the deepwater harbor closest to the goldfields and, as speculators attracted by the gold rush soon began to advocate, the likely site of a saltwater terminus for a railroad spanning the continent from east to west.

The colony of British Columbia, formed in 1858, with its capital established at New Westminster on the Fraser River in 1859, needed settlers and in 1870, a year before joining the confederation of Canada, passed the Land Ordinance Act allowing any British subject over the age of eighteen, excluding females or aboriginals, to file a preemption claim to 161 acres of crown land, whose purchase could be completed for $1.00 an acre after making improvements of at least $2.50 per acre and occupying it for four years.

John Hall, born in Kent, Ireland, in 1819, was among those who took up the offer, filing on September 22, 1870, to preempt 161 acres "situated on the North Arm of Burrard Inlet, east side, west of Deitz, Nelson, and Moody's Timber Reserve " Hall had emigrated with his older brother to Ontario, and they both came west for the Fraser River gold rush in 1859. An accomplished woodsman, he had cut trails throughout the lower mainland, staked claims to tracts on False Creek and in Coquitlam, and chose the former Tsleil-Waututh village site of Tum-Tumay-Whueton, with its southwest-facing beach and several acres of cleared, fertile land, as the place to settle down. Sewell Prescott Moody and his partners George Dietz and Hugh Nelson had secured a lease from the colonial government to 1,400 acres of forest on the north shore of Burrard Inlet, but this excluded the land around Belcarra that was the traditional home of the Tsleil-Waututh.

By the time of John Hall's arrival, most of the surviving Tsleil-Waututh had moved to Seymour Creek on the north shore of Burrard Inlet, near the growing settlement of Moodyville, leaving the Belcarra site occupied by only a handful of individuals, although that does not mean the Tsleil-Waututh, accustomed to rotating occupancy of sites within their territory, no longer believed the site belonged to them. Before he filed for preemption, Hall had warned the Tsleil-Waututh, including a man named Siamoc whom he claimed to be guilty of the murder of a white man named Crosby, and James George Sla-holt, the hereditary chief of the Tsleil-Waututh (and grandfather of the actor Chief Dan George), that "if any of these Indians complained of his pre-empting the land in question, he would have Siamoc arrested for the murder." The Tsleil-Waututh acquiesced, and Hall moved into a cabin on the site that had been occupied by one of the Indians, named Charley, and Jenny, his wife.

When the ground was dug up to plant potatoes, a large number of human skeletons were disinterred. Upon Hall's instructions, these remains were packed into wooden boxes and reburied "further out on the point."

According to Jenny, her husband "had a very sick heart regarding the removal of the bones," prompting Charley and Tom Has-yanoch, whose "friends were all buried in the ground," to join in a complaint filed by Alfred Smith, another displaced resident of the property, who had himself recently been convicted of assault for throwing a rock at Hall. Hall was charged with desecration of Indian graves and acquitted after an officer sent to investigate found Charley "kept in the house during the investigation" while Siamoc, Has-yanoch, and Sla-holt "satisfied him on the subject" despite appearing "very drunk." Hall was found guilty on a lesser charge of supplying spirituous liquors to Indians, for which he was fined thirty dollars, plus costs.[16]

Hall's preemption claim included the whole of the former settlement and much of the heavily forested Belcarra peninsula. During his field survey, completed on May 21, 22, and 23, 1874, the largest trees noted as landmarks are a cedar five feet in diameter, a hemlock three feet in diameter, and a fir two and a half feet in diameter.[17] Hall built a house and outbuildings at the far edge of the original clearing, planted a garden and an orchard on the former village site, and entered into a partnership with the hand logger Stephen Decker, his immediate neighbor living at the head of Bedwell Bay. Hall's improvements to the property were certified on May 23, 1874, and he was granted permanent title on September 4, 1882. The entire non-Native population of Burrard Inlet, including the settlements of Moodyville, Granville, Hastings, and Port Moody, was 297 at that time. Vancouver did not yet exist. Moodyville was the cultural, mercantile, and technological center of Burrard Inlet, with a library, a mechanics institute, a telegraph connection to New Westminster, and the first electric lights.

After establishing his homestead, Hall took a young woman from the Native settlement on Seymour Creek as his common-law wife. She had two children, in 1877 and 1879, and then died under unknown circumstances some time before 1882. Her mother, named Mn-maat (and known by some as Mary Dish and by others as Kate), is on record as having paddled over to Belcarra from the north side of the inlet late in the morning of October 18, 1882, with her sister Chealtah, to visit her grandchildren and to attempt to collect a debt owed to her husband by John Hall.

As daylight was fading over the inlet at the close of the afternoon, there was a confrontation over the disputed money, and John Hall shot Mary through the chest with his Winchester .44. He was charged with murder, and his tenure at Belcarra came to an end. The nature of the debt, and why Mary was there to collect it, remains unknown. J. Warren Bell, an associate of both John Hall and his neighbor Stephen Decker, would later recall that the going rate at that time was fifty dollars for a *kloochman*, or Native wife, and after that "all she would cost was her keep. You could quit her at any time or sell her but the buyer had to again pay the father or nearest relation at whatever price he asked."[18]

Who drank what, and when, were the subjects of conflicting reports. Hall's employee Peter Caulder testified that the drinking, amounting to some seven bottles of rye whiskey, had begun the day before and that he had been "introduced to a bottle at 6 o'clock in the morning" on the day of the murder by Hall. John Handcock, a neighboring homesteader, testified that he, Hall, Peter Caulder, and Stephen Decker were all at Hall's house about noon, with only half a bottle of whiskey left, when the two women, accompanied by a third Native woman, in Handcock's account, arrived in the canoe.[19]

"I took 2 drinks," Handcock testified, and "did not see the ladies take a drink at all." Handcock left about 1:30 p.m., "to put up a windlass to kill a cow," adding that "there was too much drinking for me when I have work to do." Stephen Decker left sometime between 4:00 and 5:00 p.m., later testifying that "Hall was drunk" and "Caulder not sober," and walked back home, accompanied, in his account, by a Native woman, on the seven-hundred-yard trail to Bedwell Bay.[20]

At some point, Mary began arguing with John Hall over the money, and in Chealtah's testimony "John Hall struck Mary with his fist . . . I took my sister by the arm and said, 'Better for us we go to the canoe.'" Peter Caulder testified that between 5:00 and 6:00 p.m. he "was out milking cows for Mr. Hall" when Hall's "little daughter came out and told me her father wanted me to come in." He found Hall in the doorway of the house holding on to Mary, who broke free and began to run through the garden toward the beach, with Hall in pursuit. Caulder joined the chase, tackling Mary just above the beach and retrieving a twenty-dollar gold piece from her bosom that he handed to Hall before letting her go.

Mary continued running to the beach, where Caulder caught up with her, preventing her from launching the canoe, while Hall went back to the house and returned with his sixteen-shot lever-action Winchester: a first-generation repeating rifle that still used the old .44-caliber Henry black powder rimfire cartridges favored by buffalo hunters on the Great Plains. "He had both his hands on the rifle. I won't say if his finger was on the trigger," Caulder testified. "The last I saw he had both hands on the gun." Hall was about three yards from the canoe, and Mary was stooped down beside the canoe and turning around to look at him when the gun discharged.[21]

According to the coroner's examination of the body in Moodyville on October 20, a single bullet "entered on [the] left side towards the back between the 5 & 6th ribs" and exited "between the right nipple and the breast bone," causing "almost instantaneous death."[22] According to both Caulder and Chealtah, who was watching from behind a nearby maple stump, Mary fell facedown on the beach next to the canoe, while John Hall turned around and walked calmly back up the trail through the garden to his house.

Chealtah testified through an interpreter, both at the inquest and at the trial, that she stayed hidden until dark before walking to Stephen Decker's cabin, where she found Decker and Caulder in consultation. She then returned to the beach, loaded her sister's body in the canoe, and paddled back to the Native settlement, from where word was sent to Moodyville the next day. Jonathan Miller, the sole constable for all of Burrard Inlet, started by boat for Hall's house on October 19. He found Hall, Caulder, and the two children in a canoe and arrested Hall without incident, turning him in to the magistrate in Moodyville before taking him to the one-room, wooden jail at the Granville town site, the nucleus of the future city of Vancouver, where there were two cells. Hall was charged with murder, whose penalty was death by hanging, after an inquest the following day.

The trial was held on November 27, 1882, in New Westminster before Judge Henry P. P. Crease, a friend of Hall's for more than twenty years who explained to the jury that "if I had known that John Hall would be here for trial, I assure you I would not have presided here today."[23] Crease, who had immigrated to Vancouver Island from England in 1858, "administered the law with a strong but temperate hand, often in the wildest and most distant parts of the province," and, after being gravely injured on circuit in northern British Columbia, owed his life to his First Nations companions who carried him to safety "down the steep trails not 'feet foremost like a dead man,' as he said, 'but head foremost,' by which means he persuaded them to bring him out."[24]

Hall's life was in the jury's hands. The Crown was represented by W. J. MacElmen, while the defendant was represented by William Norman Bole, a fellow Irishman who in 1877 had been on his way to Australia but missed his ship in San Francisco and took passage to Victoria, British Columbia, instead. He became the first barrister to practice in New Westminster, earning a reputation as a formidable defense lawyer before becoming queen's counsel in 1887 and a county judge of the supreme court in 1889, where he acquired the nickname "Lightning Justice" Bole. He defended a number of high-profile murder cases, including James "Scotty" Halliday, accused of the murder of Thomas Poole and his two children in 1879, who was acquitted, despite much incriminating evidence, after a trial lasting nearly a month, "mainly due to the breakdown of the principal crown witnesses under the pitiless cross examination of Mr. Bole."[25]

Bole mounted a vigorous defense of Hall, first by calling two witnesses who suggested that the gun might have discharged accidentally, although under cross-examination the first witness was forced to admit that Hall was an experienced marksman who hunted regularly, had owned this particular firearm since 1871, and carried it with him habitually in the woods. The second witness, the retired captain George Pittendrigh, a veteran of the Crimean War and commander of the British Columbia Provincial Artillery, seemed to object to repeating rifles in general but had nothing specific to add to the circumstances of the crime.

As the only hope of saving his client's life, Bole then attacked the credibility of the two eyewitnesses, first by pointing out minor inconsistencies between Peter Caulder's statement to the coroner the day after the killing and that given before the court. He then challenged Chealtah's account of the shooting, citing Stephen Decker's testimony that at the time that she claimed to have witnessed her sister's death, she had been in bed with him back at his residence on Bedwell Bay. Under cross-examination, when asked to positively identify the woman in question to the jury, Decker stumbled, saying, "They all look alike to me."[26]

"I believe that woman [Chealtah] to be a perjurer," Bole argued in his summation to the jury, advising them not "to deprive a fellow creature of his life on the testimony of a perjured witness." No white man should hang on the testimony of an Indian, he implied, though careful not to make the statement this plain. "The nature of the Indian forces him to look for vengeance," he continued. "I ask you to believe that the story told by Steve Decker is the truth, and nothing but the truth, and that the statements made by the Indian woman were prompted by the savage thirst for revenge."[27]

"A cruel murder has been perpetrated; a fellow creature was left wounded to die on the sea beach, at the North Arm," countered Mr. MacElmen for the Crown, noting that "the attempt to make the killing appear accidental cannot be sustained." He emphasized that Chealtah's description of the killing, given through an interpreter by a woman who spoke no English and did not even know how to measure time in hours by a clock, was in perfect accord with the postmortem evidence collected by the coroner, concluding that "the white people, who pretend to be Christians, acted like barbarians on this occasion, and . . . the Indians are as well entitled to protection as the whites."[28]

Judge Crease, in his final instructions to the jury, was more explicit, stating that "if you believe that woman's life was taken by John Hall, he is guilty of murder. No provocation, however great, will extenuate the act or reduce the crime to one of manslaughter." Despite his friendship with the defendant, he could find no grounds for reasonable doubt. "He walked leisurely to his own house after killing the grandmother of his children. He left her like a wild beast on the spot where she fell and looked after her no more."[29]

After two hours of deliberation, the jury delivered a verdict of manslaughter and sentenced John Hall to seven years of hard labor in the British Columbia Penitentiary. Hall escaped the gallows, only to die, shortly after his release from prison, in 1889.

Upon his engagement as counsel by John Hall, Norman Bole had visited the scene of the killing to prepare for his client's defense. As his son J. Percy Hampton Bole, born in January 1882, later described it, Norman Bole found himself "falling in love with something he could only liken to his beloved Mayo County" in Ireland, and after saving Hall from execution, in lieu of payment for services rendered, he "purchased the 160 acres, placing the balance to John Hall's credit in the Bank of British Columbia against his release." Bole named the property Belcarra, in Celtic "the fair land upon which the sun shines," after a village in county Mayo that his family had visited while he was growing up.[30]

The former Tsleil-Waututh village site, a short boat excursion from downtown Vancouver, became a favored picnic ground. Where the ancient midden was eroding into the bay, children played on the clamshell beach and dug arrowheads and spear points out of the bank. In 1904, the Boles sold the property to the Terminal Steamship Company, the first in a series of transactions that saw the original village site developed into a working-class resort with a ferry landing, a dance pavilion, picnic shelters, and overnight cabins, while the surrounding acreage was subdivided and sold to speculators as vacation lots.

Indian Arm became saturated with summer cottages, camps, and, at the extreme end of both the inlet and the socioeconomic spectrum, the Wigwam Inn, frequented by the likes of John Jacob Astor and John D. Rockefeller, nestled into the rain forest at the mouth of the Indian River like a hunting lodge on a Bavarian lake. A small fleet of passenger ferries made the trip from downtown Vancouver to Belcarra in a little over thirty minutes, faster than the trip can be made by car today. In 1908 a traveling post office and grocery began servicing Belcarra and a long list of smaller landings throughout Indian Arm.

After divesting himself of the John Hall property, Norman Bole and his wife, Florence Coulthard Bole, preempted an adjacent ninety-six acres to the immediate south. This tract was later sold by the Boles to a group of cottage owners and then repurchased in 1934 by Percy and his wife, Norah Kathleen Bole, who had already preempted an adjacent forty-six-acre parcel in 1914. Above a white sandy beach, at the southern edge of this property, they built an eleven-room craftsman-style cottage and named it the Starboard Light Lodge. A green lantern hung from the porch overlooking the inlet, with a carved wooden cradle for a shotgun just inside the door. After Percy retired in 1932, they lived there full-time, without electricity until 1955 and without a road until 1959.

As Vancouver grew into a metropolis, Burrard Inlet absorbed the associated heavy industry: sawmills, booming grounds, shingle mills, generating plants, an oil pipeline terminal, three oil refineries, grain elevators, a sugar refinery, numerous shipyards, and a barge and derrick works. The Native population was relegated to two reserves, on the North Shore, at Capilano and Seymour Creeks. Much of the unclaimed shoreline was taken up by squatters during the Great Depression, and when the Depression ended, many who had taken a liking to inlet life stayed on. In Canada, the foreshore, between high and low tide, belongs to the Crown, and squatting became the norm. Local officials complained, but other than proclaiming that the squatter community was "a nest of perverts," they could do little, because jurisdiction fell to the National Harbours Board, which generally declined to act because the shacks, although sitting on stilts within the harbor, could not be deemed a menace to navigation, because the water they obstructed dried at low tide. The last squatter living in Stanley Park was not evicted until 1958, and it was 1982 before the squatters were evicted from Admiralty Point.

Malcolm Lowry wrote most of *Under the Volcano* in a squatter's shack on the beach in Dollarton, where he lived with his wife, Margerie, from 1940 to 1954, across the inlet from the Starboard Light Lodge. He described his neighbors as a mix of seasonal residents who were "electricians, loggers, blacksmiths, mostly town-dwellers earning good salaries but not sufficient to afford summer houses at one of the settlements further up the inlet where land could be bought," interspersed with year-round residents "who had been here many many years before the summer people came, and who had their houses here by some kind of 'foreshore rights.'"[31]

"I was overwhelmed with a kind of love," he elaborated, describing his first winter on the inlet. "Standing there, in defiance of eternity, and yet as if in humble answer to it, with their weathered sidings as much a part of the natural surroundings as a Shinto temple is of the Japanese landscape, why had these shacks come to represent something to me of an indefinable goodness, even a kind of greatness? And some shadow of the truth that was later to come to me, seemed to steal over my soul, the feeling of something that man had lost, of which these shacks and cabins, brave against the elements, but at the mercy of the destroyer, were the helpless yet stalwart symbol."[32]

In 1960, Percy and Norah Kathleen Bole sold the larger of their two parcels, including the seven beachfront cottages, to Lebanese investors with interests in building an oil refinery. The new transmountain pipeline from the oil fields of Alberta terminated on the south shore of the inlet, near where George Vancouver's party had spent the night. In 1965 they sold the smaller parcel, including the lodge, to the same investors, just before Percy's death.

The Lebanese owners assigned management of the property to Montreal

Trust, whose only activity was to send an agent named Ian Worley out to the Starboard Light Lodge once a month to collect $110 in rent. The seven cabins remained in the hands of the families who by one means or another retained claim to them, while the lodge was rented first to a family named the Pearsons and, by the time the *D'Sonoqua* showed up, to the math instructor Alan Martin and the marine biologist Michael Berry, who were racing and rebuilding British sports cars in their spare time. Berry was restoring a purple wood-paneled Austin Mini, with a transverse engine and front-wheel drive, while Martin owned a Lotus Elan that was so low-slung it could pass in and out of the quarter-mile-long unpaved Bole driveway without raising the arm on the metal gate at the junction with the paved road at Bedwell Bay. In one of the outbuildings was a 1961 Riley, and in another was a vintage Humber, named Willard, used for transport into town.

In the early fall of 1971, a film industry prop and clutter director named David Hiscox chartered the *D'Sonoqua* for an afternoon cruise up Indian Arm. Hiscox, who knew everybody and could locate anything, carried a gold Dunhill cigarette lighter that Julie Christie had given him when *McCabe and Mrs. Miller*, filmed near Vancouver, wrapped. When he later dropped it overboard at the dock in False Creek, he hired a diver to get it back. As we passed Belcarra on the way back from the Wigwam Inn, Hiscox mentioned that he had been to the Starboard Light Lodge to look at a vintage MG and suggested we pay a visit, so we tied up at the ferry landing, below the former village site.

As we made our way south along the beach, traversing the face of the midden below the dance hall, then following the base of an undercut, overgrown bank in the afternoon light, our engineer Rick Cotter noticed the remains of a very old rifle among the tree roots protruding from the eroding bank. We knew nothing of the John Hall murder case at that time. The gun we found was missing its stock, encrusted with sediment, and of lever action similar to the Winchester .32-special carbine we carried on the boat.

At the Starboard Light Lodge we met Michael Berry, who gave us a tour and invited us to come back anytime. The lodge was an hour's paddle from downtown Vancouver, but a century apart.

Now, a year later, we dropped anchor in seven fathoms and launched our skiff, and I towed the log ashore. From Belcarra it was less than a mile, by water, over to the small community of Deep Cove: the end of the road from Vancouver and equipped with a store, a Laundromat, an English tea room, and the Log Cabin Restaurant, in the process of reinventing itself. The new owners were looking for some hand-split cedar shakes for the restaurant's facade. Michael Berry invited me to stay and split up the log. I left the *D'Sonoqua* and moved ashore.

To split shakes, you drive a steel blade about three inches wide and eighteen inches long, called a "froe," with a socket for a wooden handle at one end, into the end grain of the shake bolt with a maul, this being any tough piece of heavy wood—yew wood, or a piece of vine maple with a knot at one end—with one end carved down to handle size. Once a few blows from the maul have driven the froe into the bolt, you give the wooden handle a sharp pull and a shake separates with a loud snap, smooth as a sawn board and almost as flat. The best froes were made from tempered-steel leaf springs scavenged from wrecked cars.

First you have to dismantle the log into bolts. I set to work on the beach, bucking the log into twenty-four-inch rounds, then splitting the rounds radially into sections and axially into bolts. The inner core of the log is discarded, it being a matter of judgment where to make the distinction between the coarse-grained core and the fine-grained outer bolt. You are trying to identify a transition, made five or six hundred years ago, when the tree's century or more of adolescence came to an end and what we term "old growth" began. I had never split a shake in my life. The necessary unwritten knowledge was somehow either inherent within the wood itself, conveyed to me along with the tools I borrowed, or absorbed by osmosis somewhere during my travels along the coast.

Why does western red cedar split so much better than anything else?

A shake log begins as a young seedling in the forest, reaching for the light hundreds of feet overhead. Eventually, it gets there, producing its characteristic vertical grain along the way. All lateral branches are abandoned except at the top. But Douglas fir, yellow cedar, and Sitka spruce, grown in the same conditions, can be just as clear and straight yet will not split into shakes. For some reason red cedar allocates all its strength to its longitudinal fibers with almost no adhesion across the grain.

Cedar enabled dugout canoes to be built up to sixty feet in length, supporting warfare and long-distance trade. House planks could be split perfectly true and much stronger than sawn boards, enabling a form of architecture and social structure that existed nowhere else: modular longhouses that could be dismantled, packed up for transport by canoe, and moved seasonally to other sites.

Red cedar resists rot, is unpalatable to insects, and sheathes itself in thick fibrous bark that can be harvested without injuring the tree, as wool is sheared from sheep. After the bark was beaten to separate the fibers, they were woven into matting, twisted into rope, used as insulation and absorbent, or mixed with the wool of wild mountain goats and domesticated dogs to produce a soft, water-resistant equivalent of the polyester fleece we use today.

Humans and cedar flourished together on the Northwest Coast. Infants born in cedar longhouses to clans identified by carved cedar house posts were swaddled in cedar-fiber diapers, nourished by food stored in cedar boxes and served in cedar dishes the size of wheelbarrows, while growing up to wear woven cedar cloaks under woven cedar hats. After death, their remains were wrapped in cedar bark matting, packed into cedar boxes, and paddled, in cedar canoes, to burial islands where they were hoisted back into the trees, using cedar-bark ropes.

It took half the log to supply the Log Cabin Restaurant with its cosmetic facade. What about the other half? How about boards? I began to experiment and was soon splitting boards a foot wide, half an inch thick, and six feet long. I started thinking about a house. After leaving the *D'Sonoqua*, I had set up camp just south of the Bole estate, on an open granite outcrop about two hundred feet above the inlet, with a clear view out Second Narrows that was probably used as a lookout in the past. I built an open lean-to and a small fire pit but had no way to keep animals or winter out.

I was wondering how to get the freshly split boards and other building materials up to the top of the rocky bluff, and how to avoid attracting attention from the authorities, when it occurred to me that building in one of the trees I was looking down on might solve both these problems at once. Today, anyone living in a tree in British Columbia for three years is assumed to be trying to save the rain forest. The only thing I was trying to save was rent. Malcolm Lowry and his neighbors were left unmolested because their shacks fell in between the municipal authorities who had jurisdiction over structures built on land and the National Harbours Board, which had jurisdiction over structures affecting navigation by sea. Who had jurisdiction over a structure in a tree?

The Bole property was bordered to the south by the Admiralty Point Government Reserve, leased to the City of Vancouver in 1912 for a term of ninety-nine years. A small community of squatters had affixed themselves to the shoreline at Admiralty Point, despite a bronze plaque cemented into the granite headland proclaiming that "no rights have been given by the City of Vancouver to others to occupy" the reserve. At the southern edge of the former Bole property was a Douglas fir about three feet in diameter, its base only a few feet above the water at high tide. It must have been there at the time of John Hall's original survey of the shoreline and escaped the subsequent logging because its lower branches, reaching out over the water to capture unobstructed light, left little clear timber in its trunk. What made it unappealing to a hand logger made it attractive to someone looking for a place to build a house.

After reaching the lower branches, about eight inches in diameter and thirty feet above the ground, I kept climbing until I found a set of branches that could delineate the framing of a house. Then I found a better set of branches, higher up. When I decided on a location, it was ninety-five feet above the ground. I moved in at the end of November, just as a gale blew in from the north. The temperature dropped in hours and remained below freezing for two weeks. "In the morning I opened a window to look out," I reported to my father, "and the wind coming in the window blew the glass out of my newly-built door on the opposite wall."[33]

There were five windows. Two were small wood-framed octagons salvaged from a remodeled house; another two were tempered glass salvaged from the front of old television sets abandoned in a downtown Vancouver alley; and the fifth, the one that blew out, was a glass panel in the front door. One of the octagonal windows looked out across the inlet over Boulder Island toward Dollarton and the rocky beach at what was now the foot of Lowry Lane; the other looked out into the forest canopy behind the house. The plate glass window at the foot of the bed looked north over Belcarra Bay and Turtle Head toward Deep Cove and the narrow entrance to Indian Arm. The one at the head of the bed was angled downward, hinged open along the top, and overlooked the wild shoreline between the base of the tree and Admiralty Point.

I stockpiled driftwood at the base of the tree and hauled it up in heavy canvas mailbags, feeding it to a small cast-iron ship's stove, the "Trout" model, smallest of a line of stoves, all named after fish, from the Albion Stove Works in Victoria, that were common among small fishing vessels at that time. It took very little firewood to keep warm. The framing, besides the fourteen branches incorporated into the structure of the house, was made from saplings cut in the nearby forest, lashed in place with tarred nylon net mending twine. The floor was fir and spruce, and the roof and the outside walls were cedar shakes, with split cedar paneling inside. I lived there for three full years, except for summers spent traveling up and down the coast. House and contents stayed dry even when unoccupied for extended periods of time. The rain-forest floor is a wet, shaded sponge, permeated by mold and decay, while the forest canopy is well-ventilated and exposed to the light. All that rainfall, six feet and more per year, saturates the ground but passes by without stopping on its way down.

The inlet was home to thousands of wintering seabirds—grebes, surf scoters, buffleheads, harlequins, mergansers, and loons—as well as a colony of cormorants that hunted underwater by day and roosted in a nearby stand of Douglas firs at night. You could see underwater, from ninety-five feet up. The grebes dove at the same time, hunting schools of small fish trying to coordinate their escape, and then resurfaced, minutes later, as a group, somewhere else. The scoters and buffleheads dove in the near-shore shallows to graze barnacles and mussels off the rocks and then excreted the ground-up shells to be washed ashore, replenishing the white-sand pocket beaches that correlate with the flocks of ducks.

Flying squirrels were a plague. They inhabit the forest canopy, avoid the ground, and come out only at night. On the Northwest Coast, in the absence of nonhuman primates, they seem to be evolving to take the part. They would land on the roof in the middle of the night, gliding down from far up the hillside in a broad arc before pulling up at the last second to flare their landing, claws out. They adopted the house as a cache for Douglas fir cones, returning as the cones ripened to peel them apart like artichokes, eating the seeds concealed inside, as well as anything edible not secured inside metal or glass. They either already knew, or quickly learned, how to unscrew the tops off jars. I built a live trap and took dozens of captives over to the other side of the inlet or to Boulder Island, from where they could not glide back.

The tree house was sheltered from the prevailing southeast winter gales but exposed to the southwesterlies that swept in with a change in fronts. In gusts above fifty or sixty knots, the tree would reduce its windage by letting a few branches break off cleanly at the trunk. The motion of a boat is periodic, a superposition of the natural period of the boat and the period of the surrounding waves. The motion of the tree house was aperiodic, with a deflection of as much as ten or twelve feet in storms, following a chaotic trajectory that never described the same path twice. From the top of the tree you could see over the Belcarra peninsula to the mountains above Indian Arm, and when winter brought arctic outflow winds, they were funneled toward the tree house by the fjord. You could pour a cup of tea out the window, and it would freeze before reaching the ground. The same path that had been taken by a lobe of the Cordilleran ice sheet, stripping away the landscape down to bedrock, was now being taken by the arctic air that stripped branches off the trees as it passed by.

Living without telephone, computer, internet, or even electric light, I had time beyond measure to think. I found myself thinking about what, if anything, a tree might think. Not thinking the way we think, but the way a single neuron thinks, integrating information over time. It might take years to register the premonition of an idea, centuries for an entire forest, networked through synapses established by chemical signaling pathways among its roots, to form a thought. After three years I was no closer to an understanding, except to have gained a lingering suspicion that trees were, in some real and tangible way, as John Ambrose Fleming put it, "manifested Thoughts in a Universal Mind."[34]

I was caught between two worlds: surrounded by rain forest, with the sea below, yet looking out Burrard Inlet toward downtown Vancouver, I saw construction cranes appear on the horizon, leaving high-rise apartment buildings behind them as they spread. In the winter darkness, the headlights of commuters heading home in the late afternoon on the Barnet Highway slowed to a crawl. When the air was calm, the report of the nine o'clock gun, a relic of a time when ships at anchor set their chronometers the evening before departure, still arrived at forty seconds past nine from Brockton Point in Stanley Park.

Growth rings in trees are Nature's way of digitizing time. Some of the split cedar boards paneling the walls of the tree house spanned seven hundred years. I counted the grain in one seven-inch board, and it went back to the year 1426. Halfway through that board, in 1679, Leibniz had imagined his digital computer, with marbles running along mechanical tracks. Two and a half inches ago, in 1778, James Cook had arrived on the Northwest Coast. Bering and Chirikov had arrived half an inch earlier, in 1741. My entire life, so far, spanned one-quarter of an inch.

6.

STRING THEORY

The Stone Age is termed the Stone Age not because stone tools were the most advanced technology in the Paleolithic tool kit but because they are the best preserved. Lashing, sewing, and weaving were as sophisticated as the fabrication of edge tools, but only an indirect record has endured.

Tying things together is as important as cutting them apart. A knot can be as useful as a knife, a net can be as effective as a spear. The tensile arts were not so much invented as adopted: from vines hanging in the forest, sinews exposed when animals were dismembered, hides stretched out when they were skinned, beach grass weaving the idea of basketry in the wind. Net making is as old as the spider's web.

The introduction of bifacial, fluted stone points, culminating in the Clovis tradition that flourished across North America at the close of the Pleistocene, was as much an advance in the art of hafting these points to the weapons that delivered them as it was an advance in the working of

Mount S! Elias.

View when Mount S! Elias bore N. W.b W. 20 lea.*

the stone itself. It was lashing and sewing, as much as fire and flint, that enabled the human species to span the earth.

Just occasionally you find yourself in an odd situation," begins Thor Heyerdahl's account of a voyage that took place in 1947, inspired by his field research on the British Columbia coast. "You get into it by degrees and in the most natural way but, when you are right in the midst of it, you are suddenly astonished and ask yourself how in the world it all came about."[1] Heyerdahl then described how he came to find himself in the middle of the Pacific Ocean, with five companions, on a raft lashed together from South American balsa logs, sailing forty-three hundred nautical miles to Polynesia, where they made landfall after a voyage of 101 days. *The* Kon-Tiki *Expedition* was the first grown-up book I read.

I was eight years old, with two younger sisters, when a new babysitter named Helen Dukas showed up. Helen, born in Germany in 1896, had grown up in Freiburg as one of seven children, left school to help take care of her siblings after her mother died in 1909, became a kindergarten teacher in 1919, and accompanied Albert Einstein to Princeton in 1933, serving as his personal secretary and household manager until his death in 1955. To the world, Einstein achieved immortality through his work. To his friends and neighbors in Princeton, Einstein achieved immortality through Helen Dukas. She was his literary executor, archivist, and companion to his stepdaughter Margot until Helen's death in 1982. There were no children in the Einstein household, a deficiency we were chosen by Helen to fill.

One winter afternoon, too cold to play outside, I was bouncing around on the large green spring-loaded Naugahyde recliner in my father's study, making Helen's life difficult, when she suggested, "Why don't you *read* something?" I answered, "There's nothing to read!" The study was full of books, but not children's books. Helen went to the shelf, picked out a book, and said, "This one. Read this!"

The book was *The* Kon-Tiki *Expedition*, by Heyerdahl. I was transported out of the green recliner into the balsa forests of Peru, where logs felled by hand were floated down the Palenque and Guayas Rivers to the sea, lashed together under the bewildered eye of the Peruvian navy, then towed out of the harbor to catch the Humboldt current and the trade winds to the islands of the South Seas. By the third sentence I learned that on the morning of May 17, 1947, Heyerdahl had "found seven flying fish on deck, one squid on the cabin roof, and one unknown fish in Torstein's sleeping bag."[2] By the fourth photo caption I learned that "we lashed the nine balsa logs together with ordinary hemp ropes, using neither nails nor metal in any form."

After I read *The* Kon-Tiki *Expedition*, all I wanted to do was lash things together and build boats. When I needed a shortwave radio antenna, I lashed together a forty-foot mast from bamboo that grew along the edge of the Oppenheimers' fence. When I built my first kayak skeleton, I followed the instructions to use screws, but followed up with lashings where they split the kiln-dried wood. When I took up rock climbing, I was more interested in knots, ropes, and slings than in the rock itself. When I took up sailing, I was more interested in the running rigging than the sails. When I needed a place to live, I lashed together the framework of a tree house in a Douglas fir. When I split the back of my right hand open with an ax, I sewed it up, using dental floss and a sail needle, with my left hand. I never saw a structural problem that lashing or sewing could not solve.

Einstein remembered being five years old, and sick in bed, when he was given a small magnetic compass, leading him to begin puzzling over the nature of electromagnetic fields. I, too, was given a small magnetic compass, but it failed to turn me into a physicist. I knew it could be counted on to point north and south by aligning itself with the earth's magnetic field, but I did not wonder, "*Why* are there magnetic fields?"

In 1948, the United Nations Educational, Scientific, and Cultural Organization issued a small booklet, *Suggestions for Science Teachers in the Devastated Countries*, expanded and reissued as *700 Science Experiments for Everyone* in 1958. I learned more tinkering my way through these experiments, requiring only scavenged or discarded materials, than I ever did in school. The problem with the otherwise excellent educational system in Princeton, New Jersey, was the assumption of a lifelong distinction between manual trades and intellectual skills. Helen Dukas disagreed.

Heyerdahl left Norway in September 1939 with his wife, Liv, and their infant son, arriving in Vancouver just as he turned twenty-five. He had an older cousin, Jens Conrad Heyerdahl, who had settled on Vancouver Island in the 1920s; very little money, a folding kayak, and a conviction that similarities in material culture between the Northwest Coast and Polynesia were evidence of early seafarers having made the voyage from east to west. "He said he didn't have any money to buy gas," remembers Clayton Mack, of Bella Coola on the central B.C. coast, who offered the Norwegian visitor the use of his own vessel to search for petroglyphs to support his case. "He told me that he was broke. The German people took all his money when they raided Norway. Took all the money out of the banks." Mack paid for the gas himself. "During the day we would go look at them Indian paintings on the rocks. At night we would sit around and drink our coffee and talk. Heyerdahl told me Hawaiian people are Bella Coola people. 'Bullshit,' I said. They can't go across the ocean. 'I'll make my own canoe,' he said. 'I'll go across someday.'"[3]

Heyerdahl might have sailed a dugout canoe instead of a balsa raft to Polynesia, had not a call to join the free Norwegian forces in the liberation of Finnmark from the Nazis intervened. When the war was over, a tenuous theory of pre-Columbian Nordic voyages to Mesoamerica brought him to Peru, rather than back to the Northwest Coast. Heyerdahl's theory of Polynesian settlement is now discredited, but this has not diminished what he and the crew of the *Kon-Tiki* did.

When Europeans arrived on the Northwest Coast of America in the eighteenth century, they were met by two distinct populations: those who built skin boats and those who built dugout canoes. The skin-boat builders assembled bits and pieces of driftwood into a framework that formed the minimal, skeletal delineation of a boat. The dugout builders started

with a single tree trunk, removing everything that did not fit their definition of a canoe. The dugout was a creature of the rain forest. The skin boat
was a creature of the sea.

The two species of boatbuilding, before becoming separated by geography, were separated in time. Skin-boat builders appear to have arrived in
advance of the mature forests required for dugout canoes. The boundaries
between them have always been shifting: over the long term driven by
the retreat of glaciers and the advance of forests, and over the short term
by incursions and cultural appropriation from both sides. There were areas of overlap in between, such as Prince William Sound, where Chugach
skin-boat builders occupied a heavily forested coastline without taking up
dugout canoe building, and Yakutat, where Tlingit dugout canoe builders
adopted kayaks from the Chugach for limited use.

The arrival of the Russians brought this uneasy balance to an end.
Prevailing colonial practice, in the eighteenth century, was to extinguish
indigenous technology, disrupting the associated culture as its material basis was transformed. The Russians, in contrast, adopted the Aleut kayak,
renamed "baidarka," as the primary vehicle for their settlement of the
Aleutians and their subsequent expansion deep into dugout canoe territory
along the rain-forested Northwest Coast.

After Steller and his fellow survivors of the shipwreck of the *St. Peter*
returned to Kamchatka, the merchant Emel'ian Basov sponsored the construction of a small vessel, the *Sv. (Saint) Apostol Peter,* taking both the
name of its predecessor and a number of Bering's former crewmen on a
return voyage to Bering Island to collect furs. They left on August 1, 1743,
wintered on Bering Island, and returned the following year with a catch
sufficient for the crew to donate fifty otter skins to the Church of St. Nicholas, "the patron of northern hunters, fishermen, and sailors," in Nizhne-
Kamchatsk.[4] The *Sv. Apostol Peter* sailed again in 1745, venturing farther
eastward to the "Unknown Islands" and returning with sixteen hundred sea
otter skins. Three more voyages by the *Sv. Apostol Peter* followed, spawning

a succession of private ventures that swept through the Aleutian Islands to Kodiak Island and Prince William Sound.

Sea otters (*Enhydra lutris*) are insulated not by a layer of blubber like other northern sea mammals but by a coat of thick luxurious fur. They had been hunted to near extinction on the Asiatic coast, and their pelts, favored by Chinese emperors, fetched high prices in the markets of Kiakhta and Canton. British and American traders began arriving on the Northwest Coast in the 1780s, their ships stocked with goods, from needles to axes to firearms, that were traded for furs, then taken to China and exchanged for tea and silk before the voyage home. The Russians, with little or nothing to trade, had to capture the furs themselves, enlisting the locals, at times by force, to help.

The *Sv. Apostol Peter* and its siblings, built on the Kamchatkan coast, were known as *shitiki*, constructed from small planks of wood sewn together along their edges with lashings of sea mammal skin, and sometimes equipped, in the absence of canvas, with reindeer-skin sails. The lineage of these vessels can be traced back to Slav traders of the tenth century, based in Novgorod, making contact with Norse invaders who introduced their Viking-era boatbuilding and seafaring skills. They were a mongrel hybrid between a planked vessel and a skin boat, crewed by a mixed lot of Russians, Kamchadals, and part Kamchadals.

The *promyshlenniki* who led these expeditions were descended from the Pomory who lived on the shores of the White Sea: boatbuilders, walrus hunters, and traders in fur and ivory who migrated eastward through Siberia, pioneering the arctic shipping routes and settling on the Kamchatkan coast. "The ways of life and appearance and manners of people they encountered in Alaska were neither novel nor strange to them," explains Lydia Black, the Russian American historian and ethnographer who devoted much of her life to illuminating this complex period through a mix of archival and field research. "Though they fought with the Aleuts and in the end conquered them, there was much they shared."[5] They followed practices established in Siberia and Kamchatka, forcing the Americans to pay *iasak*, or tribute, in furs. Upon their return to Kamchatka, one-tenth of the take was claimed by the government, which received its share ahead of the church.

Catherine the Great, who became empress in 1762, took a personal interest in the new islands and their inhabitants, issuing orders to the navigators not to abuse her new subjects and to report on everything they found. These instructions were often ignored. "I have read with satisfaction your account of the discovery of six hitherto unknown Aleutian Islands which have been brought under my scepter," Catherine wrote to Denis Chicherin, the governor of Siberia, in 1766. "This discovery is most pleasing to me." She thanked Chicherin for the artifacts sent to St. Petersburg, noting that "the bags woven of grass, the thread made of fish gut and the bone hooks are very skillfully made."

"I am sending you a muff which was sewn here from one of the skins you sent; this may serve as an example of how the skins should be prepared," she continued, adding that "local women of fashion wear flounces and edging on their garments made of furs. It would be a good idea if you were to select some, especially those that are colorful, and send them here. If you do not know what flounces are, ask your wife, and she will inform you."

Catherine instructed Chicherin to "have one of the local natives brought back, but give orders that no one is to be brought unwillingly or by any force whatsoever, only someone who is willing to come," adding a postscript in the margin: "Instruct the promyshlenniks to treat the inhabitants of the new islands with kindness and not to deceive or mistreat them when they collect iasak from them."[6]

Reports of abuses against the local inhabitants and incursions by foreign traders prompted Catherine, in 1786, to order the deployment of a naval squadron, under the command of Grigorii Ivanovich Mulovskii, from the Baltic to the Aleutians and Northwest Coast, both to reassert possession and to investigate these reports. Four battleships and an armed transport were commissioned, along with 34 officers and a contingent of scientists, forming a total crew of 639. To avoid the miseries of Bering's voyage, antiscorbutics and desalinization equipment were requisitioned, as well as special clothing, including twelve pairs of socks for each person on board.

Mulovskii, while authorized to destroy any foreign settlements or shipping found encroaching on Russian territory anywhere between Vancouver Island and the Aleutians, was forbidden to use force against Native inhabitants or even to retaliate if attacked. A separate overland expedition under Joseph Billings, who had served aboard the *Discovery* during Cook's third voyage and later offered his services to the Russians, was to depart from Okhotsk and Kamchatka to rendezvous in Alaska with Mulovskii's fleet.[7]

In 1787, Sweden declared war on Russia, forcing Catherine to cancel Mulovskii's orders and deploy his ships against the Swedish fleet instead. Billings, who had already left St. Petersburg in 1785 in command of the "Northeastern Secret Geographical and Astronomical Expedition," was left to complete his instructions as best he could. He was twenty-four years old and his second-in-command, Gavriil A. Sarychev, was twenty-two. They retraced the overland journey taken by the Bering-Chirikov expedition, building two ships, the *Slava Rossii* (Glory of Russia) and the *Dobroe Namerenie* (Good Intent) in Okhotsk. The *Dobroe Namerenie* was wrecked in September 1789 upon leaving Okhotsk and was burned to recover its iron. The *Slava Rossii* alone sailed for Kamchatka, where its crew spent the winter, arriving in the Aleutians, at Unalaska, on June 1, 1790. They were met by Aleut paddlers "sporting about more like amphibious animals than human beings," as Martin Sauer, the expedition's secretary, noted, adding that "they row with ease, in a sea moderately smooth, about ten miles in the hour, and they keep the sea in a fresh gale of wind."[8]

"The head of the boat is double the lower part, sharp, and the upper flat, resembling the open mouth of a fish," Sauer observed, "and they tie a stick from one to the other to prevent its entangling with the sea weeds,"[9] echoing James Cook, who had described the kayaks he observed at Unalaska in 1778 as terminating "in a forked point, the upper point projecting without the under one which is even with the Surface of the Water," adding that "why they should construct them in this manner is difficult to conceive, as the fork is liable to catch hold of everything that comes in the way."[10]

The Aleuts had discovered bow-wave phase-cancellation effects, whose rediscovery would lead to the divided, bulbous bows seen on all oil tankers and most large cargo vessels today. A moving object pushing down on the surface, like the upper bow of a ship, produces a wave that begins with a crest, whereas a moving object pushing up on the surface, like a protruding lower bow, produces a wave that begins with a trough. The two wave systems cancel each other out, making the vessel faster and more efficient because of the reduced wave-making resistance, and also quieter and less detectable to its prey.

The open-jawed, bird-tailed vessels that attracted notice at the time of first contact soon disappeared. Other than a small handful of specimens collected by the early Russian round-the-world voyages and rumors of precontact kayak skeletons surviving in undocumented burial caves, few examples exist. To the Russians, the high-speed kayak, whose design had been driven not only by the pursuit of sea mammals but also by warfare between different island groups, was a threat. It enabled Aleut warriors to coordinate a series of uprisings against the initial Russian occupation, taking advantage of superior speed, stealth, and mobility in the same way that Apaches on foot or mounted on light war ponies had outmaneuvered the U.S. Cavalry sent in pursuit.

Under Russian administration, sea otter hunting shifted to fleets of larger kayaks, roaming farther afield. Conflicts between rival groups of hunters, concern over unreported or underreported revenue, and intense lobbying by Grigorii Shelikhov, an entrepreneur later investigated, under Catherine's instructions, as a war criminal for his massacre of noncombatants who had retreated to safety at Refuge Rock at Kodiak in 1784, led to the establishment of the Russian-American Company in 1799. Modeled on the East India Company and the Hudson's Bay Company, the Russian-American

Company was a government-sanctioned monopoly with complete control over both the fur harvest and the kayak builders whose skills made it possible to capture the animals that were wearing the furs.

Between 1741 and 1867, when Russian America was sold to the United States, an estimated 264,800 sea otters were captured by Russian sea-otter-hunting expeditions in Alaska and California.[11] The hunting parties, fielding fleets of as many as 750 baidarkas, escalated the long-standing conflict between the kayak builders and the dugout canoe builders, with the Aleut and Alutiiq paddlers using strength in numbers and Russian firearms to invade Chugach and Tlingit territory, plundering all the sea otters that could be found. It was long-standing enmity against the kayak builders, as well as anger at the misbehavior of the Russians, that led the Tlingits to destroy the Russian settlements at Sitka in 1802 and at Yakutat in 1805.

Those making the annual voyages across the Gulf of Alaska, with hundred-mile stretches without harbors and seas that break a full two miles offshore, faced a choice between death by drowning at sea or death at the hands of the Tlingits (Kolosh in Russian nomenclature) if they sought shelter ashore. "Many such paddlers, once they have made two such journeys, fall into a fever and are completely without strength," noted Gavriil Davydov, who survived extensive kayak travels in Alaska, only to drown, at age twenty-five, in the Neva River in St. Petersburg, after trying to jump across an opening drawbridge after a night of celebration with his partner Nikolai Khvostov in 1809.[12] At other times the paddlers had the strength to continue to safety, but their boats fell apart. The "List of Fatal Mishaps Which Befell Kodiak Aleuts in the Early Years of Baranov's Stay," compiled by Kyrill Khlebnikov, includes 135 paddlers who were "poisoned by shellfish" returning from Sitka in 1799; 35 who drowned in 1800; 165 "killed en route to Sitka, by Kolosh," in 1802; 16 "killed by Kolosh during attack on Sitka" in 1804; and 100 "drowned en route to Kodiak from Sitka" in 1805, as well as 200 "drowned in baidarkas during storms in that same year."[13]

———

The Russians extended the baidarka in length as well as range. To carry passengers, cargo, and weapons up to the size of small swivel guns and two-pound falconets, they commissioned the construction of three-hatch baidarkas up to thirty feet long. The first detailed description appears in the journals of Cook's third voyage, on October 14, 1778, at Unalaska, when a Russian navigator, Gerasim Grigor'evich Izmailov, then thirty-three, sought out the British ships. "He enterd the Bay where the Town stands in great State," noted Charles Clerke, "being attended by 30 or 40 Canoes, all of which however were the Single Man Canoes, excepting it with the European Cargo; it was paddled by two Men and had a third Hole, formed for the accommodation of his Russian Honour . . . Every Individual of this little Fleet seem'd totally devoted to the Service of this russian Gentleman, and were no sooner landed than great part of them were busied in his Employ, some building a temporary House for his Nights repose, others dressing Fish different ways for his Evening regale, and rendering various other little Services, to make matters as agreeable to him as circumstances cou'd possibly admit of "[14]

Cook found Izmailov to be "a very sencible intelligent man" and "very well acquainted with the Geography of these parts." Invited aboard the *Resolution*, he "at once pointed out the errors in the Modern Maps."[15] It was Izmailov who, six years later, in command of the *Tri Sviatitelia* (Three Saints) would help Shelikhov subjugate the Kodiak Islanders and establish a permanent settlement, before exploring and charting the mainland coastline to Yakutat, Lituya Bay, and beyond. Questioned under oath during the Billings-Sarychev inquest, he admitted killing unarmed prisoners, but only under direct orders from Shelikhov.[16] He led numerous hunting expeditions and in 1787 returned to Kamchatka with a record-breaking 172,000 rubles' worth of furs.

"This Mr. Ismyloff seemed to have abilities to intitle him to a higher station in life than that in which he was employed," Cook concluded after visits were exchanged. "He was tolerably well versed in Astronomy and other necessary parts of the Mathematicks. I complemented him with an Hadlys Octant and tho it was the first he had perhaps ever seen, yet he made himself acquainted with most of the uses that Instrument is capable of in a very short time."[17] Thomas Edgar, master of the *Discovery*, concurred: "This said Ishmiloff is a native of Jacutz & is a very shrew'd penetrating sort of a man, his Ideas never confused and his conception vastly quick."[18]

The three-hatch baidarka became the light utility vehicle of the Russian-American enterprise, transporting supplies, harvested furs, company personnel, visiting dignitaries, medical practitioners, and Russian Orthodox priests carrying elaborate portable churches for delivering services in outlying settlements, equipped with roll-up icons that could fit inside the boat. A long list of visitors to Alaska adopted the three-hatch craft. "In my opinion," wrote the German physician and adventurer Georg Heinrich von Langsdorff in 1814, after extensive travels in Alaska and California, "these baidarkas are the best means yet discovered by mankind to go from place to place."[19]

The people of the Aleutian Islands, having had the archipelago to themselves for ten thousand years, suffered not only the arrival of the Russians but also a series of aftershocks that followed. Catherine II, seeking to stabilize the new colonies, legitimized the children of Russian-Native marriages, who were recognized as free Russian citizens and entitled to education and other benefits, including an exemption from military service, at Russian-American Company expense. Within two or three generations much of the day-to-day administration and technical operations in Russian America were in the hands of Native-born "creoles."

The Russian Orthodox Church transliterated the scriptures via the Cyrillic alphabet into indigenous languages that might otherwise not have been preserved. Ivan Veniaminov, head priest of the Unalaska District, later metropolitan of Moscow, and, since 1977, a saint, compiled the first systematic meteorological records in Alaska, an Aleut-Russian dictionary, and the first comprehensive ethnography of the Aleuts, from information gathered firsthand. "It seems to me that the Aleut baidarka is so perfect of its kind," he concluded, "that even a mathematician could add very little if anything to improve its seaworthy qualities."[20]

Overhunting led to a catastrophic decline in fur seal and sea otter populations, until the Russian-American Company instituted a sea mammal protection program and the number of animals began to rebound. Other businesses were cultivated, such as coal mining on the Kenai Peninsula, the export of ice from Kodiak to California, the casting of church bells for the California missions out of native copper, and the publication of an atlas of the North Pacific printed in St. Petersburg from copper plates engraved in Sitka by the creole mechanic Koz'ma Terent'ev, following charts drawn by the company navigator Mikhail Kadin, also a creole.

Sitka, founded in 1799 as Novo Arkhangel'sk, had surpassed Kodiak to become the capital of Russian America, with a library, hospital, museum, multiple churches, and schools. Except for a token deployment of Siberian guards in a small barracks in Sitka, there was no military presence in the Russian American colonies other than the irregular visits of Russian naval vessels making round-the-world tours. In 1812 a small settlement, Fort Ross, was established on the coast of California, just north of San Francisco, both to serve as a base for sea-otter-hunting parties and to supply the Alaskan settlements with fruit and grain.

The Crimean War of 1853–1856, pitting Russia against an alliance of England and France, led Great Britain to expand and fortify the Hudson's Bay Company's base of operations at Victoria on Vancouver Island. When gold was discovered in the interior of British Columbia in 1858, the trickle of settlers became a flood. As it came time to renew the Russian-American Company's charter in 1862, there was grumbling in high places over the costs of maintaining the American settlements against the encroachment of the Hudson's Bay Company, and a decision was reached, without consulting the people living in Alaska, to sell the Russian American territory and its fixed assets for $7.2 million to the United States.

Few Alaskans accepted the offer of transport "home" to a Russia that had never been their home. The creole population, numbering over two thousand, lost their privileged status overnight. "The Creoles are unfit to exercise franchise, as American citizens," William H. Dall, staff naturalist at the Smithsonian Institution and one of the first to survey the new territory, advised.[21] The rights of the Native population were left unresolved until the Alaska Native Claims Settlement Act of 1971.

At four o'clock in the afternoon on October 18, 1867, before an assembled crowd including the Russian-American Company's chief manager, Prince Maksutov, who, with his brother, had gained fame during the Crimean War by defending Petropavlovsk against attack by the French and British fleet, the Russian flag was lowered over Sitka, until it was snagged by one of the yardarms and had to be cut free by a Russian sailor before falling the rest of the way to the ground. The U.S. flag was then raised to a salute from the *Ossipee*, the *John L. Stephens*, the *Resaca*, and the *Jamestown*, U.S. Navy warships anchored in Sitka Bay.

The U.S. generals Jefferson Davis and Henry W. Halleck were sent to establish a temporary government, having distinguished themselves in the Civil War. Because fish and game were plentiful and settlers were scarce, they managed to avoid the conflicts that had led to the recent Indian Wars in the western United States. In his initial census of the population, Halleck classified the "Esquimaux" as "low in the scale of humanity, and generally harmless, but often treacherous and hostile to small parties of whites," and the Kenaians as a "proud and fearless race, peaceful and well disposed," but "ready to avenge any affront or wrong." The Aleuts he found "generally kind and well disposed, and not entirely wanting in industry," while warning that although they "now pretend to be friendly," the Tlingits "are all of so treacherous a character that depredations by them may be expected on the first favorable opportunity."[22]

There was now an influx of itinerant non-Native sea otter hunters, with the catch reaching a high of 4,152 in 1885 before a precipitous decline. The hunt was then limited to Native hunters using traditional methods before being outlawed entirely in 1911. The next upheaval was World War II. The Aleutian archipelago became a battleground in determining whether the United States would establish bases from which to attack Japan, or Japan would establish bases from which to attack the United States. The inhabitants of the contested islands were either captured by the Japanese or evacuated by the United States to safety in Southeast Alaska, where they were housed in overcrowded cannery bunkhouses at places like Excursion Inlet, under conditions little different from internment camps.

The construction of wartime airfields left the pattern of settlement permanently changed. The last remnants of the kayak fleets, having survived until the war, succumbed to the aircraft, outboard engines, and gasoline that followed. William S. Laughlin, who had toured the Aleutians in 1938 with Aleš Hrdlička, collecting mummies from burial caves for the Smithsonian, returned a decade later to the village of Nikolski on Umnak Island, an island that had once supported thirty-five large settlements, and removed the last two baidarkas that were left.[23]

According to Laughlin, it was Steve Bezezekoff who squeezed the dynamometer so hard that it broke when Waldemar Jochelson and Dina Brodsky were testing grip strength at Nikolski in 1910. "When people do that much kayaking it should show on their skeletons, and it does," Laughlin explains. "Be assured there is nothing as massive, nor as rugose, as the Aleut male kayak-hunter's humerus," he says, citing a specimen excavated in 1950 as "an example of maximum development within the human species, both for what the Germans called *massigkeit*, and for the size and rugosity of individual muscle insertions, or origins."[24] These multiple, elongated insertions are indicators of extreme stamina, achieved through redundancy that allows individual muscles to escape fatigue without the paddler's stopping to rest, the way that migratory birds are able to fly for days nonstop.

The Aleut kayak's bifurcated bow attracted the most attention, but there was just as much genius in the design of its compound, truncate stern, evolved to facilitate planing at high speed and minimize the quarter wave, which often exacts more of a penalty than the bow wave in a high-speed craft. Waves, and wave-making resistance, are produced both when the front of a vessel parts the water and then again when the vessel's hull turns inward at the "quarter" to rejoin at the stern. The baidarka's flared, high-volume stern both minimizes the quarter wave and produces dynamic lift when surfing in following seas.

More puzzling are the effects of a spring-loaded, non-dissipative articulated skeleton and the nonlinear elasticity of sea mammal skin. Some baidarka skeletons, fastened with elastic, low-friction baleen-fiber lashings, incorporated intricate ivory ball-and-socket joints and contained as many as sixty of these moving parts, perhaps allowing some of the opposing wave energy to pass unimpeded through the skeleton, without being absorbed by the kayak and its paddler as physical work.

Fifteen years after the last baidarka left Nikolski, I was twelve years old, living in New Jersey, and began building a crude wood-framed kayak, oblivious to the ten thousand years of accumulated knowledge that was evaporating, at that very moment, four thousand miles away. I could have gone to Nikolski, Unalaska, or Atka and learned from the last of the master builders who were still alive at the time, instead of stumbling around on my own.

Five years later, we launched the *D'Sonoqua* and headed up the British Columbia coast, most of it less populated than it had been in precontact times. We spent the first winter in Quatsino Sound at the northwest end of Vancouver Island, where immigrants from Norway, England, Finland, Ireland, and Denmark, along with refugees from eastern Canada and the prairies, had established homesteads and settled down. We cut two masts from a grove of Sitka spruce near Mahatta River that belonged to Charles Bland, whose wife, Lilian, had designed, built, and flown her own airplane in Ireland before moving to Canada and whose daughters Mary and Dora, both artists, had lived there their entire lives. We field dressed the masts before towing them farther up the inlet, where we boxed them down to heartwood in a boat shed belonging to a former Norwegian homestead

Fleet of "over 500 baidarkas, with two or three men in a boat," according to its commander, Egor Purtov, as encountered at Port Dick, near Cook Inlet, by the Vancouver expedition on May 16, 1794. (Engraving after a sketch drawn "on the spot" by Harry Humphrys, aboard the *Chatham*, from George Vancouver, *A Voyage of Discovery to the North Pacific Ocean, and Round the World*, 1798)

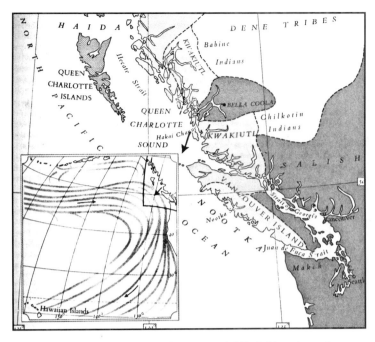

Detail from a chart based on Thor Heyerdahl's fieldwork on the coast of British Columbia in 1940. The arrow pointing southwest from Hakai Passage indicates a theorized point of departure for voyages by canoe to Polynesia. (After Thor Heyerdahl, *American Indians in the Pacific*, 1952)

Clayton Mack and Thor Heyerdahl in a folding kayak with a black bear, Kwatna Inlet, British Columbia, 1940. (Photograph by Cliff Kopas, courtesy of Joe Ziner)

Kwakwaka'wakw (Kwakiutl) dugout canoes under sail off Deer Island in Queen Charlotte Strait near Fort Rupert, British Columbia, during the filming of Edward Curtis's *In the Land of the Head Hunters*, 1914. (Photograph by Edward S. Curtis, courtesy of Charles Deering McCormick Library of Special Collections, Northwestern University)

"Frame of a canoe found at Port des Français. By the side is the covering of Seal Skins which served instead of Planks." Engraving, after a sketch drawn at Lituya Bay in 1786, showing a vessel belonging to a trading party from the north. Mount Fairweather (15,300 feet) appears in the background. (J.F.G. de La Pérouse, *A Voyage Round the World*, London, 1799)

George Dyson with an aluminum tubing kayak frame, Vancouver, British Columbia, 1971, constructed after a Nunivak Island design collected in 1889. (Photograph by Ron Orieux)

LEFT George Dyson aboard the harbor tug *Widgeon*, tied up next to a 110-foot chip barge at Double Bay, Hanson Island, British Columbia, 1973. (Photograph by Ron Keller)

RIGHT The *Widgeon* and tow nearing Juneau, Alaska, after a thousand-mile voyage, at a top speed of four miles per hour, through the Inside Passage from Tacoma, Washington, August 1973.

Nicholas II, the last tsar of Russia, with Stanislaw Neverovsky, captain of the Imperial Yacht *Standart*, on a baidarka excursion off the coast of Finland, 1912. (Photograph by Count Alexander Grabbe, courtesy Paul Grabbe)

Бaйдapka однолючная Остpoвa уналашки

One-hatch baidarka from Unalaska, drawn by James Shields, ca. 1798, showing the bifurcated, surface-piercing bow. (Photocopied by F. A. Golder in St. Petersburg in 1914, courtesy of University of Washington Libraries, Special Collections, UW 1770)

Three-hatch Chugach baidarka, built by Stepan Britskalov, in front of the chapel at Chenega, Prince William Sound, 1933. (Photograph by Frederica de Laguna, courtesy of Frederica de Laguna and the Department of Ethnography, National Museum of Denmark)

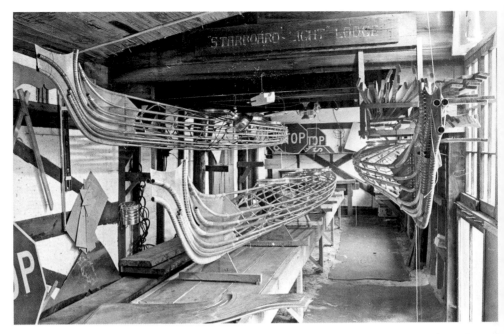

Three-hatch baidarkas under construction at the Starboard Light Lodge, March 1977.

Three-hatch baidarkas under tow in the Gulf of Alaska off Chichagof Island, May 1977.

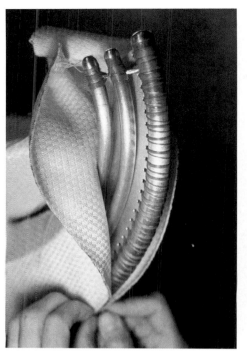

ABOVE Sewing a nylon skin covering on an aluminum-tubing kayak frame. (Photograph by Ann Yow)

LEFT Lauren and George Dyson with baidarka skeleton, Bellingham, Washington, 1993. (Photograph by Ann Yow)

BELOW The mystery is not so much how the early form of the Aleut kayak with a surface-piercing lower bow developed, but why it suddenly died out.

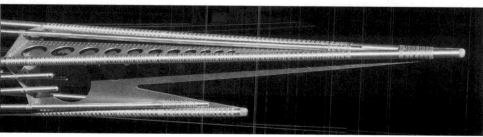

Sailing south, off the face of the La Perouse glacier. Gulf of Alaska, June 1977.

Joe Ziner at the reins, surfing downwind in Clarence Strait, Southeast Alaska, July 1977.

Crossing Dixon Entrance between Southeast Alaska and British Columbia, August 1977.

Camp, with upturned sails for shelter, Admiralty Island, Southeast Alaska, July 1977.

LEFT Cutting floorboards for the *Mount Fairweather*, Hanson Island, British Columbia, 1974. A Stihl 090 chainsaw with a forty-eight-inch bar and an "Alaskan Mill" attachment, designed by Will Malloff of neighboring Swanson Island, is being used to saw the forty-foot boards out of a windfallen Sitka spruce. (Photograph by Ron Keller)

ABOVE Bobbi Innes and Robert Hunter (Greenpeace) helping with the construction of the *Mount Fairweather*, 1975. (Photograph by Rex Weyler)

Launching of the *Mount Fairweather*, Belcarra Park, British Columbia, June 21, 1975. (Photograph by Peter Thomas)

Interior of the *Mount Fairweather* under construction, March 1975. (Photograph by Peter Thomas)

Leaving Hanson Island, head ng south from Blackfish Sound into Johnstone Strait, August 1975. (Photograph by Paul Spong)

Bishop Samuel Butler (1774–1839), Charles Darwin's headmaster at Shrewsbury School, on the speciation of languages, 1833. "If any body likes to write an 8vo on these hints he may," notes Butler, explaining with a diagram how languages descending from a common parent diverge.
(Courtesy of Shropshire Archives, Shrewsbury, UK)

ABOVE Samuel Butler (1835–1902) as an undergraduate at Cambridge, 1858, before leaving for New Zealand. (Lithograph from a photograph taken at St. John's College, after Henry Festing Jones, ed., *A First Year in Canterbury Settlement*, 1923)

RIGHT Opening paragraph of Samuel Butler's letter to the editor of the Christchurch *Press*, June 13, 1863. (Courtesy of Alexander Turnbull Library, New Zealand)

Correspondence.

DARWIN AMONG THE MACHINES.

TO THE EDITOR OF THE PRESS.

SIR,—There are few things of which the present generation is more justly proud than of the wonderful improvements which are daily taking place in all sorts of mechanical appliances. And indeed it is matter for great congratulation on many grounds. It is unnecessary to mention these here, for they are sufficiently obvious; our present business lies with considerations which may somewhat tend to humble our pride, and to make us think seriously of the future prospects of the human race. If we revert to the earliest primordial types of mechanical life, to the lever, the wedge, the inclined plane, the screw, and the pulley, or (for analogy would lead us one step further) to that one primordial type from which all the mechanical kingdom has been developed, we mean to the lever itself, and if we then examine the machinery of the Great Eastern, we find ourselves almost awestruck at the vast development of the mechanical world, at the gigantic strides with which it has advanced in comparison with the slow progress of the animal and vegetable kingdoms. We shall find it impossible to refrain from asking ourselves what the end of this mighty movement is to be. In what direction is it tending? What will be its upshot? To give a few imperfect hints towards the solution of these questions is the object of the present letter.

Mesopotamia, Samuel Butler's homestead at the headwaters of the Rangitata in New Zealand, where he wrote "Darwin Among the Machines" in 1863, below the alpine valley that would become the setting for *Erewhon; or, Over the Range* in 1872. (Photograph by Edward Percy Sealy, May 1866, courtesy of Michael Graham-Stewart)

Samuel Butler's sheep brand (*left*) and branding iron (*right*), registered November 26, 1860, to his Mesopotamia run. (Print after H. F. Jones, 1823; photograph by permission of the Master and Fellows of St. John's College, Cambridge)

Thomas Huxley to Charles Darwin, February 3, 1880. After Samuel Butler criticized Darwin over his failure to acknowledge prior evolutionists, Huxley came to Darwin's defense: "Has Mivart bitten him and given him Darwinophobia? . . . It's a horrid disease and I would kill any son of a [sketch] I found running loose with it without mercy." (Courtesy of Cambridge University Library)

Wovoka (Jack Wilson), a Paiute prophet of the Ghost Dance movement, seated with his uncle Charley Sheep, outside his reed-thatched dwelling in the Mason Valley near Walker Lake, Nevada, after a heavy snowfall, January 2, 1892. (Photograph by James Mooney, courtesy of National Anthropological Archives, Smithsonian Institution, INV 06285500)

Arapaho Ghost shirt collected by James Mooney in 1890. (After James Mooney, 1896)

Burial of Sioux casualties by U.S. Army contractors, January 3, 1891, at Wounded Knee. "The bodies of the women and children were found scattered along for 2 miles from the scene of the encounter, showing that they had been killed while trying to escape," reported James Mooney, who included this charcoal drawing, after a photograph, in his account. (After James Mooney, 1896)

Analog revisited. Seventy years after analog electronic components were assembled into digital computers in the aftermath of World War II, digital computers are being assembled into analog systems that treat streams of bits as continuous functions, the way streams of electrons were once treated in vacuum tubes, like this twin-triode 6J6.

Time is; Time was; Time is past. Friar Bacon and Friar Bungey, thirteenth-century pioneers of artificial intelligence, have fallen asleep, while their assistant Miles plays music and fails to wake them when the brass head they have so laboriously constructed finally speaks. Woodcut from *The Famous History of Frier Bacon*, 1679. (Courtesy of Huntington Library, San Marino, California, RB 111363)

in Hecate cove. Below the main house was a one-cylinder drag saw for cutting up beach logs, hauled into position by a dedicated windlass, then split and stored under the house below a trapdoor into the kitchen next to the stove. Out back was a gas-powered wringer washing machine that was started like a motorcycle by giving a pedal a few sharp kicks.

British Columbia was dugout canoe territory, but a kayak was the answer to turning the minimum of material into the maximum of boat. When we returned to Vancouver in the spring, tying up at the National Harbours Board fishermen's terminal in False Creek, I began building another kayak, lofting its lines from a sketch of a Nunivak Island kayak in Edwin Tappan Adney and Howard Chapelle's *Bark Canoes and Skin Boats of North America*, published by the Smithsonian Institution in 1964 and sold for $3.75. I was attracted by its trademark open hole in the bow, its handgrip tail at the stern, and a large hatchway, which turned out, along with its unstable hull form, to be unsuited to the unskilled.

I made the frame out of aluminum tubing, purchased from Wilkinson metals at the end of Fraser Street on the North Arm of the Fraser River and carried back to my sister's house in Kitsilano lashed to the frame of my bike. I could pedal about two blocks before it started oscillating from side to side, forcing me to stop until all trace of oscillation disappeared before being able to pedal another two blocks. I formed the ribs and stringers by bending the tubing over a scrap-metal jig, then lashed everything together with nylon seine twine purchased from Redden Net down the street. When you lash two pieces of high-tempered, thin-wall aluminum tubing together with eight turns of 60-pound test twine, you can count on it taking 8 x 4 x 60, or 1,920 pounds, to pull them apart. Lashings distribute the load equally, and there is no risk of metal fatigue or losing the temper at a weld. When you examine how nature solves high-stress attachments— mending a broken bone, joining the plates in an infant's skull, healing a cut in an animal's skin, or attaching a mussel's shell to a surf-pounded rock— you find fibrous lashings carrying the load.

It was luck, not skill, that kept me afloat. Instead of learning the missing skills, I built a larger more stable boat. In *The Bark Canoes and Skin*

Boats of North America, Howard Chapelle mentioned large three-hatch baidarkas having been used in Alaska, but he considered them an invasive species introduced by the Russians and outside the scope of his book. He did include a "Kayak from Kodiak Island, 2-hole Aleutian type," from the collection of the Washington State Historical Society and Museum, its lines having been taken off by John Heath in 1962. I bought some more aluminum tubing and lofted out its lines, adding a third hatchway and a dragon-headed prow, and extended the length to thirty-one feet.

Heyerdahl's construction of the *Kon-Tiki* confirmed the ability of a balsa raft to cross the Pacific Ocean, and the construction of a three-hatch baidarka confirmed why the Russians had taken such a liking to this design. It was stable enough to stand up in, was fast under paddle, and could carry half a ton. The limit to its capacity was less how much stuff could be stowed away inside it and more how willing you were to carry all that stuff up and down a slippery beach when making or breaking camp. I launched it in August 1972, paddled around the remainder of that summer, added two small triangular lug-rigged sails that winter—my first winter in the tree house—and in early April 1973 headed north, hitching a tow from the *D'Sonoqua* as far as Desolation Sound. I had no approved life jacket, no radio, no running lights, and no GPS, only a hand compass and a few outdated paper charts.

At Desolation Sound the tidal currents divide, flowing southward to return to the Pacific via the Salish Sea and northward through a maze of channels to reach the Pacific through Johnstone Strait. Among the constricted northern passages, the currents reach speeds of twelve knots and follow baffling patterns, with the water flowing in even after the level has been falling for up to two hours and then continuing to flow out with a rising tide. "The irregularity of the tides was such that no correct inferences could well be drawn," complained Vancouver in 1792. "The time of high water was equally vague and undefinable."[25] In July 1792, four weeks after their joint survey of Burrard Inlet, Vancouver and Broughton managed to take the *Discovery* and the *Chatham* through what is now known as Seymour Narrows and Discovery Passage, while the Spaniards Galiano and Valdés, in the smaller schooners *Sutil* and *Mexicana*, attempted the narrower passage to the north, now known as Cordero Channel, including Dent Rapids, "the place the English considered the worst."[26]

Galiano and Valdés, scouting the entrance to Arran Rapids with their launch, were met by locals returning from harvesting cedar bark who, along with "the first fresh salmon we had seen in the strait, and a great quantity of sardines . . . gave them to understand that they ought not to risk passing through the channel in the longboat for they would be swamped beyond help in the whirlpools," warning them further against attempting the passage in the schooners, "deducing that the size and resistance of our vessels would not promise us a more happy lot, but rather a more disastrous one than they had met with their canoes." After waiting for slack water, the two schooners made it through, experiencing "such violent whirlpools that the water sank more than a yard," with the *Sutil* "turned round thrice with such violence that those on board were made giddy," and "rendering it impossible to steer." They spent the night in a "terrifying" anchorage behind Dent Island where "the violent rapidity of the waters in the channel caused a frightful noise and a most noticeable surf."[27]

We tied up the *D'Sonoqua* at Heriot Bay on Quadra Island on April 17, where my logbook notes that I "visited Pub but met no one wishing to canoe through Johnstone Straits." I left the *D'Sonoqua* the following day and paddled north, entering the Yuculta Rapids at sunset, with a full moon, on the first of the ebb, the current strengthening as I turned west into Gillard Passage, where the chart notes that "violent eddies and whirlpools exist." I then coasted downhill toward Dent Rapids and Devil's Hole, where the chart warns that "violent eddies and whirlpools form." By now the ebb was really picking up, and this was my chance to do what I had always imagined when passing through in larger boats: spend the night on one of the islets across from Devil's Hole.

Whirlpools emerged at intervals throughout the night, like a toilet flushing in an airport restroom by itself. Seals and otters hauled themselves out on the rocks a few yards away and crunched on the bones and exoskeletons of what they had caught. The whirlpools appeared with no warning from the depths, only to disappear and then re-form. Large high-floating logs were sucked out of sight, resurfacing in calm water somewhere else.

At first light I caught a back eddy on the last of the flood and continued into Johnstone Strait, where a favorable southeast wind picked up, filling the sails as the kayak surfed downwind, making more than twelve knots in the following seas. I spent the summer exploring Queen Charlotte Strait and Queen Charlotte Sound, venturing out into the open sea as far as Cape Caution, with Paul Spong's growing facilities at Hanson Island as home base. Returning from Rivers Inlet in August, I stopped in at the Nimpkish Hotel in Alert Bay, where an out-of-place character named Carroll Martin was in the beer parlor awaiting a replacement water pump for the engine on a small tugboat that had broken down in Blackfish Sound. He had tied up his tug and barge to some pilings in Double Bay, around the corner from Spong's place, and hitched a ride into town.

Martin was a rancher, elk-hunting guide, and former school principal from Bayfield, Colorado, near Durango, who had decided to sell out and move north. With his nine-year-old son as deckhand, he was towing a 110-foot retired Foss chip barge to Juneau, Alaska, behind a 36-foot tug. Dwarfed by its tow, the *Widgeon* was a World War II harbor tug with a GM 6–71 engine and had no towing winch, no anchor, no lifelines, no radio, no radar, no depth sounder, and no instruments besides a single compass. Its hull was planked in Alaskan yellow cedar, now returning to its birthplace in the north. Martin, who had fished Southeast Alaska on a seiner one summer, had purchased the tug and barge in Tacoma, Washington, before trucking his belongings, including three head of cattle and four quarter horses, up from Colorado. Before departure, twenty-five tons of Pasco valley hay had been delivered to the barge.

The 6–71, flagship of the Detroit Diesel Division of General Motors, was the six-cylinder in-line diesel that drove America into the second half of the twentieth century, powering everything from city buses to D-day landing craft to Sherman tanks. In principle you can tow the *Queen Mary* with an outboard skiff, but you won't get anywhere against wind or tide. At full throttle in a flat calm, Martin's outfit made just under four knots. This was paddling speed, so everything I had learned about tides and currents was just what was needed to help the *Widgeon* make its way up the coast. Martin and the water pump hitched a ride back to Double Bay, where he fixed the engine, while I went to retrieve my belongings and loaded the three-hatch baidarka on the barge. It was Noah's Ark.

If you spend enough time on the Inside Passage, sooner or later you see everything imaginable, and some things unimaginable, pass by. We were pushing the edge of that scale. "You probably are interested in the horses," I wrote to my younger sisters, who now numbered four and were starting to ride horses at that time. "I don't know much about them except that I haul a lot of water for them to drink. One is a Palomino, and there is a Sorrel mare with a four-month-old colt, and the fourth horse is just plain brown as far as I know. The cattle are pretty well behaved except for the white cow which got loose and fell off the barge and was a very tough job to get back up."[28]

We crawled north at a walking pace. A chip barge has a ten-foot-high timber wall around it, and Martin had chainsawed an opening at one end to load the livestock, a full-sized house trailer, a pickup truck, the twenty-five tons of hay, a speedboat, and enough lumber to build a house. Outside that wall, you were on the Northwest Coast, with the wind, the waves, the morning fog, the extended sunsets, and the periodic noise of flocks of seabirds that were undisturbed by the tug and barge's slow, quiet approach until we nudged up to them and they took flight. We were navigating without instruments, but moving so slowly that we did not worry about being holed by floating logs and had ample time to plot our course.

Inside the wall, you were in Colorado. You could see the sky overhead, but you couldn't see over the walls of the barge and might have been walking around a corral in Durango on an overcast afternoon. Once a day we stopped towing and tied up to the barge to do the chores. The livestock had to be fed and watered and their stables cleaned, giving us a chance to stretch our legs after being crammed into the tugboat, with its small wheelhouse and four rudimentary bunks within six feet of the engine running at full throttle day and night. Martin had left a six-inch layer of wood chips on the deck of the barge, which blended in well with the stables, behind a metal fence, and the pile of hay at one end of the barge, covered by tarps. It got colder and wetter as we made our way north, yet when you broke open a fresh bale of hay the smell of eastern Washington in midsummer wafted out.

Sometimes the two worlds collided, such as when we brought the barge as gently as possible alongside one of the cannery docks in Petersburg, but still with an impact that shook the dock so strongly that the cannery workers came running out. Or in Angoon, where we were greeted by a crowd of children who had never seen a cow or a horse. While waiting for slack tide to go through Wrangell Narrows, we had to anchor the barge, which had a two-hundred-pound old navy anchor on deck, attached to a coil of wire rope—but with no way of hoisting it back up. Martin, who had thought of everything, had left a clear path along one side of the barge deck and the work-around to the anchor winch problem was to drive the pickup truck, in low gear, back and forth across the deck, while his deckhands gained another eighty feet on raising the anchor by tying off the anchor cable and coiling up the slack while he backed up for another pass. Besides his son, whose younger siblings were already in Juneau with their mother, we had a sixteen-year-old hitchhiker from Tucson picked up on the drive to Tacoma, another hitchhiker picked up among the Gulf Islands in Canada, and me.

A chest freezer full of elk meat was lashed on deck. Martin did most of his elk hunting in the fall in the high country above Durango up Vallecito Creek, which happened to have been the trailhead for the last Sierra Club base camp I had worked on, after the moon landing, in 1969. After we had closed up camp at the beginning of September and were hiking out for the last time, we passed a party of elk hunters on horseback heading up the trail. Now we had crossed paths again. Day after day, in between standing wheel watches looking out over the top of the freezer, we ate elk.

After three weeks of adventures and misadventures, including running out of fuel at the mouth of Tracy Arm, we arrived in Juneau and unloaded the livestock, their trailers, and the truck. Then came the most harrowing part of the voyage. Martin retrieved a second truck, already in Juneau with his wife, handed me the keys, and told me to follow him to where the livestock were going to be stabled on some pastureland he had rented among the beach meadows just north of town. I told him I didn't have my driver's license. "This is Alaska," he answered. "You don't need your license!" How to explain that I had never learned how to drive? To a Colorado rancher, this was as incomprehensible as if I had revealed I was from Mars.

After getting the livestock settled, we secured the barge to some pilings in the middle of Gastineau Channel and tied up the *Widgeon* in the city boat harbor to rest. I spent a month working for Martin building winter stalls for the animals, sleeping on the barge in a small apartment excavated in the remaining pile of hay. It was like a hidden chamber in an Egyptian pyramid and as dry as Pasco, Washington, on a July afternoon. You could lie down in clothes wet from working in the rain, and within minutes they dried out.

In late September, I collected three hundred dollars for the work on the stables and paddled north from Juneau, heading for Glacier Bay via Point Retreat and Lynn Canal. I rounded Point Couverden into Icy Strait, which lived up to its name to the extent that with clear skies frost was beginning to settle on my sleeping bag at night, but it was no longer the spectacle it had been in August 1804, when the ice sheet that had filled Glacier

Bay was in full retreat, and Alexander Baranov, rounding Cape Spencer aboard the *Ermak*, accompanied by a party of three hundred baidarkas, described entering Icy Strait "like going into the mouth of Hell! Among icebergs which were like mountains and touched the yards!"[29]

As I paddled along the shore between Point Gustavus and Bartlett Cove, at the mouth of Glacier Bay, I saw people on the beach and coasted in. It was a group of high school students from Juneau on a field trip with their science teacher, Charles Jurasz, known for discovering bubble-net feeding by humpback whales, which he had observed releasing circular curtains of bubbles to concentrate schools of small fish down to a size they can then scoop up. When I told Mr. Jurasz that a group of humpbacks had approached the beach near Point Gustavus where I had slept the previous night and begun vocalizing through their blowholes, answered by a group of wolves onshore, he did not seem surprised. He invited me to Bartlett Cove, where the National Park Service had its headquarters, and introduced me to Greg Streveler, the park service's lead biologist, who not only had studied the life and journals of Georg Wilhelm Steller and the rest of the Bering-Chirikov expedition but also had a deep knowledge of how the Russian American adventure had unfolded, beginning with the Russian interactions with the Hoonah Tlingits who controlled the territory where Chirikov's crew had disappeared in 1741.

Glacier Bay National Monument, not yet a national park, incorporated a sixty-mile stretch of the outer coast, from Cape Spencer to Cape Fairweather, including Lituya Bay, where in 1958 a giant wave had removed the mature forest on one side of the bay to an elevation of 1,670 feet. The outer coast, guarded by the face of the La Perouse glacier extending out into the open ocean, was fringed by the largest yellow cedar groves on record and populated by brown bears weighing up to half a ton. Streveler was organizing a group of scientists to operate out of a remote inlet named Torch Bay the next season to collect baseline data for an environmental impact study triggered by a proposal for a nickel mine under the ice field that loomed above the outer coast. I mentioned my experience working for Sierra Club backcountry base camps. Would I be interested in a job as camp cook for the summer? Yes!

Streveler drew up a contract stipulating that I would show up for duty in May 1974, staying until the weather locked as if it were time to head south. The Jurasz family, the Strevelers, and Bob Howe, the superintendent, took me under their wing, arranging for me to store the thirty-one-foot kayak for the winter among the rangers' cabins in Bartlett Cove. The lodge operating under a park concession in Bartlett Cove was closing up for the season, and its tour boat was heading south. I hitched a ride as far as Ketchikan and then, on a Bering Sea crab boat, to Seattle from there.

I returned to Glacier Bay in April and paddled out Icy Strait and around Cape Spencer to Torch Bay. There were still snowbanks on some of the beaches and winter-sized Gulf of Alaska swells. Almost everything within hiking or paddling distance of Torch Bay was unexplored. Kenneth Brower, who had begun writing a *New Yorker* profile that would later develop into his 1978 book, *The Starship and the Canoe*, came out to the camp to visit, under a program that encouraged journalists to visit national parks, and joined me in paddling south. Streveler sent out his two-volume set of the journals of Steller, Bering, and Chirikov, and I studied every word of their accounts.

Mount Fairweather towered overhead, and beyond the mouth of Torch Bay stretched the open Gulf of Alaska coast, toward the narrow, tide-swept entrance to Lituya Bay, where twenty-six French seamen belonging to the La Pérouse expedition had drowned in 1786. I looked out at the open Gulf of Alaska from the windswept beach and imagined the fleets of hundreds of baidarkas that had once passed by, not that long ago, when the mouth of Glacier Bay was still an unbroken wall of ice.

In 1973, I had begun planning a much larger boat, intending to double the thirty-one-foot baidarka to sixty-two feet in length. I submitted plans for

"An Ocean-Going Canoe for Canada's West Coast" to the Canada Council's Explorations Program, asking for twelve hundred dollars in funding,
and to the Aluminum Company of Canada, asking for twelve hundred feet
of aluminum tubing. Both said no.

When I returned to Hanson Island after the summer in Alaska, I started
work on the huge canoe but scaled it down to forty-six feet. With help
from Will Malloff on Swanson Island and Paul Spong on Hanson Island,
I milled a windfallen Sitka spruce into boards that were forty feet long to
line the skeleton of the new boat. Using the money from the park service,
I bought enough tubing to build the *Mount Fairweather*, launched on June
21, 1975. A forty-six-foot kayak did not make much sense. It might have
enabled a small group of people, led by a Thor Heyerdahl, to escape the
Northwest Coast with enough food and water to make it to Polynesia, but
there were good reasons, besides the lack of aluminum tubing, that the
Russians had extended the Aleut baidarka to about thirty feet and then
stopped.

After drifting around in the *Mount Fairweather* for two years, I decided to build a small fleet of six three-hatch baidarkas, reconstructing a
1 percent sample of one of the large fleets of Russian American times. In
early May 1977, the six baidarkas, built at the Starboard Light Lodge in
a frantic four months, including making our own paddles, rain gear, life
jackets, and sails, were taken to Alaska aboard the *Betty L.*, a retired East
Coast halibut schooner turned West Coast fish packer, and launched into
the Gulf of Alaska on the outer coast of Chichagof Island, close to where
Chirikov's party had vanished in 1741.

Building one of these vessels, you translate your own strength in pulling the lashings tight into the structural integrity of the boat. It acquires a
life of its own. The resulting baidarkas, twenty-eight feet in length, were
fast under paddle and flew downwind with dragon-headed bows and fan-
shaped sails. They were stable enough to carry sail in gale-force winds,
large enough to sleep in, when necessary, tied up in a kelp bed while
waiting for a tide to turn or anchored until the wind died down, and safe

enough to keep paddling in, if needed, all night. When we did go ashore, we slept under lean-tos made from the upturned sails.

The last of the six baidarkas returned two years later paddled by people I had never met. I gave up trying to organize expeditions, or build larger and more innovative vessels, and began trying to better understand why the early baidarkas had been able to achieve their reported speeds under paddle, by building modern versions of the ancient designs. I picked up just enough mathematics and hydrodynamics to make sense of wave-making resistance, dissipative versus non-dissipative hull oscillation, and skin friction along flexible walls under turbulent flow.

An object moving though a fluid suffers a sudden increase in resistance when the boundary layer changes from laminar to turbulent flow. Most efforts at drag reduction have focused on delaying this transition, but unless the moving object is able to swim like a fish and avoid separation of the boundary layer, it can be delayed only so far. An open question is to what extent sea mammals, and possibly kayaks covered in sea mammal skin, are able to not only delay the transition but also actively or passively dampen the effects of turbulence past the point to which it can be delayed.

Those who wonder how people limited to "primitive" technology could have developed such a sophisticated design, or doubt that they could have understood how it worked even if they had developed it, have never paddled at night. The most powerful tool in modern hydrodynamics, whether for designing kayaks or supersonic aircraft, is computational fluid dynamics: three-dimensional models of fluid flow, with the pressure and velocity of every element of an arbitrarily fine computational mesh rendered visible to the designer trying to understand the flow.

The North Pacific, in summer, blooms with plankton that emit pulses of light when disturbed or subjected to a sudden pressure change of any kind. Every paddle stroke reveals the complex of whorls and eddies shed by the paddle blade; every passing fish or sea mammal reveals a trail of oscillating pressure gradients; paddling alongside a sister kayak, you see turbulence and other signs of wasted energy illuminated in real time. Only now, with tools that can duplicate a bioluminescent fluid, is our understanding catching up.

The Aleut kayak, an optimization between maximum speed, minimum resistance, and seaworthiness, occupied a summit on a complex fitness landscape that cannot be reached by climbing one of the gradients at a time. Once the summit is reached, the design tends to stay there, but it remains an open question in evolutionary theory whether by running the experiment again the same summit would be found.

The Aleuts regarded their kayaks as fellow living beings: not lifelike, but alive. Every element of the vessel was imbued with life, from the skeleton to the skin to the ballast stones, analogous to the stones found in sea lion stomachs, that were distributed fore and aft to tune the oscillation of the kayak among waves. The kayak and its builder spent both life and death together, sharing a place in the burial cave.

"What are organisms?" asked Carl Woese, the evolutionary biologist who long argued, against prevailing dogma, that the tree of life is not a tree and that we needed to add a third kingdom of life. "Imagine a child playing in a woodland stream, poking a stick into an eddy in the flowing current, thereby disrupting it. But the eddy quickly reforms. The child disperses it again. Again it reforms, and the fascinating game goes on. There you have it! Organisms are resilient patterns in a turbulent flow—patterns in an energy flow."[30]

These eddies persist as the individual elements they are composed of move on, the way the whirlpools in Cordero Channel reappear, though the water they are composed of is different from one minute to the next. The same is true for vessels: the Tlingit canoe, the Aleut kayak, and the Polynesian outrigger are morphological eddies that persisted for millennia, independent of the individuals building them. The Polynesian culture that Heyerdahl was so determined to explain the source of is latent in the landscape and seascape itself. Give the South Pacific atolls enough time, and sailing canoes will form, whether they arrive from the east or from the west. In the North Pacific, dugout canoes are latent where there are forests, and skin boats are latent where there aren't.

My rough approximation to the morphological eddy of the Aleutian kayak appeared, briefly, framed in aluminum tubing and enveloped in a synthetic skin. One day this eddy may return once again framed in driftwood and enveloped in sea mammal skin, and another day it may return as a fully synthetic yet living vessel, once the boundaries fall entirely between organisms that are grown and structures that are built. The Aleuts got there first.

When Thor Heyerdahl developed his theories about early navigation in the Pacific, he focused on the Northwest Coast, finding parallels between the Kwakiutl canoe builders of northern Vancouver Island and the Maori canoe builders in New Zealand, with Polynesia in between. "Kwakiutl Islanders, bred on the very sea-coast of the roaring Pacific, and raised from childhood with a canoe designed and used for fishing in the deep sea," he noted, "lived as near to Hawaii as Hawaii was to Tahiti."[31]

Among the 821 pages of *American Indians in the Pacific*, based on Heyerdahl's early work but published only after the *Kon-Tiki* voyage gained him fame, Heyerdahl includes a map of the Northwest Coast and its canoe-building populations, with an inset showing the prevailing ocean currents sweeping in a great arc along the Northwest Coast and offshore toward Hawaii and the Polynesian islands beyond. On this map he drew a large blue arrow, fifty miles long, extending from the north end of Calvert Island toward the Scott Islands off the northwest end of Vancouver Island, to mark the departure point for the voyage to Hawaii he had in mind.

The area around Calvert Island, Goose Island, and Hakai Passage, where Heyerdahl placed this arrow, happens to be a "hinge point," where sea level following the last glacial maximum, about fifteen thousand years ago, was close to where it is today. The average sea level was much lower than today, but where that sea level met the local topography varied widely according to the land's response to the melting ice.

Two nearby sites evidencing early human occupation have recently been found: on Calvert Island, including a series of human footprints that date to thirteen thousand years ago, and on Triquet Island, including evidence of sea mammal hunting, also going back thirteen thousand years. No direct evidence of skin-boat building has yet been identified, but indirect evidence suggests seagoing vessels and sea mammal hunters on the Northwest Coast before the forests had matured to dugout-canoe-building size. They would have had the entire coast to themselves.

There were times I wondered, like Heyerdahl, where I was and how it had all come about. One night in late August 1976, I was sailing the *Mount Fairweather* alone down Johnstone Strait, with a stiff northwesterly, a wind that usually dies down at sunset, but had picked up as it got dark. There was a sockeye opening, with myriad small gillnetters setting their nets at random here and there in the strait, and the occasional fish packer lying at anchor in one of the indentations along the shoreline waiting to pick up the catch. As the canoe gained speed in the phosphorescent water, its translucent hull glowed with a pale blue light.

The *Mount Fairweather*, skimming along at twelve knots, carried no running lights and was invisible except for the flat, intersecting aluminum plates in its bow structure that formed a corner-wave reflector giving it the radar signature of a much larger boat. To the gillnet fleet, it looked as if a large, unlit vessel were plowing down the strait. Gillnets are marked by lights at the ends, in those days still kerosene lanterns left burning on small wooden floats, so the maze of nets was difficult to make out. We drew only four inches of water and slid harmlessly over the cork lines, the gillnetters scanning in our direction with their searchlights, trying to identify the rogue vessel that was running through their nets.

As we passed the Broken Islands, the stars sweeping an arc across the August sky, we left the sockeye opening behind and had the strait to ourselves. For thousands of years, perhaps back to the end of the Pleistocene, when the narrow channel between the mainland and Vancouver Island last opened up, people had glided over the waves through the wind tunnel of Johnstone Strait.

Between the mountains lay a trail of eddies through time, lingering in our wake before diffusing among the stars.

7

EREWHON REVISITED

Radicals are increasing in each language unknown to the other, while still there is a certain common stock—but the common stock never increases," explained Samuel Butler, the headmaster of Shrewsbury and bishop of Lichfield, on May 24, 1833. Butler enforced rigid discipline among his pupils and adhered to the doctrines of the Anglican church. In his spare time he puzzled over the speciation of languages, observing that "the number of words not common to both languages therefore continually and rapidly increases, while that of words common to both remains the same." This variation, he conjectured, drives the emergence of new languages, leaving traces between languages descending from a common parent many steps removed. "And thus I think we may see how we get two languages, with some traits of resemblance and more of discrepancy," he concluded, adding, "This is all I know about it—and if any body likes to write an 8vo on these hints he may—for there has been many a one written on scantier material."[1]

Mount Edgcumbe

View when Mount Edgcumbe bore N W.b N 8 leaˢ distˤ

On the Origin of Species, an octavo volume of 502 pages, was published by Charles Darwin, Butler's most famous student, in 1859. Darwin never mentioned his headmaster's influence except to note, in his autobiography, that "nothing could have been worse for the development of my mind than Dr. Butler's school."[2] More influence was acknowledged in the other direction, with Butler's son, Canon Thomas Butler, rector of Langar and an undergraduate colleague of Darwin's at Cambridge, noting that Darwin "inoculated me with a taste for Botany which has stuck by me all my life."[3] Relations between the Butlers and the Darwins then took a turn for the worse. Thomas Butler's first son, named Samuel after his grandfather in 1835, became Darwin's most persistent critic, questioning the assumptions of Darwinism with the same skepticism he brought to the orthodoxies of his father's church. "Even a man of genius could isolate himself by antagonizing Darwin on the one hand and the Church on the other," observed George Bernard Shaw in the introduction to his *Back to Methuselah*, a "metabiological pentateuch" based on Butler's ideas.[4]

Young Samuel Butler, taking up as a novelist, proved an embarrassment to his family, with the shock of his being unveiled as the author of the anonymous *Erewhon* blamed for hastening his mother's death. He left his autobiographical, anti-Victorian masterpiece, *The Way of All Flesh*, unpublished to spare his relatives further distress. Upon his graduation from Cambridge University in 1858, Butler had been determined to become an artist, but his father refused to allow him to join "a set of men who as a class do not bear the highest character for morality [and] are thrown into the midst of the most serious temptations . . . I can't consent to it."[5]

Canon Butler gave his son a choice between the law, the teaching profession, and the church—or leaving England. After an unhappy six months in London doing unpaid parish work in preparation for being ordained, Samuel Butler considered emigrating to both South Africa and British Columbia before deciding on Canterbury Settlement in New Zealand, founded in 1850 by Church of England pilgrims who had been granted "waste land in the middle island" suitable for sheep farming and little else. Canon Butler provided an introduction to the archbishop of Canterbury, president of the Canterbury Association, and released the first installment of young Samuel's share in his grandfather's estate, relieved that his problematic son was heading as far from England as it was possible to get.

Butler booked passage to New Zealand on the ship *Burmah*, but when a consignment of livestock encroached upon the passenger accommodations, he canceled his booking to wait for the next ship. "No one in his right mind will go second class," he advised, "if he can, by any hook or crook, raise money enough to go first."[6] The *Burmah* was last seen, among icebergs, at 48° south on November 17 and then disappeared without a trace. Butler sailed instead aboard the *Roman Emperor* on October 1, 1859, just as *Origin of Species* was going to press. Saloon-class passengers embarked on the eve of departure at Gravesend; steerage-class passengers had already boarded at the East India docks. "Having clambered over the ship's side and found myself on deck," Butler reported, "I was somewhat taken aback with the apparently inextricable confusion of everything on board; the slush upon the decks, the crying, the kissing, the mustering of the passengers, the stowing away of baggage still left upon the decks."[7] A cold rain was falling, and most of the emigrants would never see England again. No one from Butler's family came to bid farewell. Samuel Butler settled into his cabin and went to sleep without saying his prayers for the first time in his life.

The *Roman Emperor* raised anchor at dawn and was delivered by a steam-powered tug to the mouth of the Thames. They made their way out of the English Channel under sail, anchoring to await fair tide off Ramsgate and then again at Deal before their last ties to England were severed. For six weeks they sailed south. Approaching the equator, they found themselves becalmed, with mysterious currents welling up from the depths as "the empty flour-casks drifted ahead of us and to one side." In the third week of November, as the first copies of *Origin of Species* were taken up by the booksellers of London, the *Roman Emperor* entered the Southern Hemisphere with "the sun vertically overhead," the would-be painter observing that he cast no shadow on the deck. The winds of the Southern Ocean soon lifted sails and spirits alike. "Huge albatrosses, molimorks (a smaller albatross), Cape hens, Cape pigeons, parsons, boobies, whale birds, mutton birds, and many more, wheel continually about the ship's stern," Butler noted as they approached the Cape of Good Hope.[8]

"The world begins to feel very small when one finds one can get half round it in three months," he reported upon arrival at Port Lyttelton, near Christchurch, on January 27, 1860.[9] The twenty-four-year-old neophyte invested fifty-five pounds in an experienced horse named Doctor, "a good river-horse, and very strong."[10] Together they explored the surrounding territory, wandering so far afield that at least once Butler was mistaken for a Maori, before settling upon a subalpine tract at the divided headwaters of the Rangitata River, where he put up a mud-walled hut, filed claim to six thousand acres of rangeland, stocked it with some three thousand sheep, and lived in comfortable isolation until selling out at a gain of thirty-six hundred pounds in 1864. He christened his homestead Mesopotamia and branded his sheep with the image of a candlestick in silhouette.

Although distant from civilization, Butler was not a recluse and enjoyed the company of his hired hands and the close-knit society of Christchurch when he descended for intellectual refreshment and supplies. "I shall never forget the small dark man with the penetrating eyes," remembered Sir Joshua Strange Williams, then a lawyer and later a justice of the supreme court, "who took up a run at the back of beyond, carted a piano up there on a bullock dray, and passed his solitary evenings playing Bach's fugues; and who, when he emerged from his solitude and came down to Christchurch, was the most fascinating of companions."[11] Robert Booth, who hired on with Butler during Mesopotamia's second year, remembered him as "a literary man, and his snug sitting-room was fitted with books and easy chairs—a piano also . . . Butler, Cook, and I would repair to the sitting-room, and round a glorious fire smoked or read or listened to Butler's piano. It was the most civilised experience I had had of up-country life."[12]

Butler's nearest neighbors, Charles and Ellen Tripp, lived eighteen miles away below the gorge of the Rangitata at Mount Peel. Ellen Tripp, daughter of Henry John Harper, the first bishop of Canterbury, found Butler of "a peculiar nature, and full of wild theories," and "did not like it when he tried to convert our maid to his ideas."[13] Ellen Tripp had been confirmed, before leaving England for New Zealand, by Bishop Samuel Wilberforce, who would later ask Thomas Huxley, standing in for Charles Darwin in a public debate, whether he was descended from an ape on his mother's or his father's side. During the first rainy season at their remote outpost, she had studied William Paley's *Natural Theology*, professing the argument from design that Darwin's theory would undermine, noting that it was one of the "few dry books we possessed." She found Butler's views "very upsetting," but he "played the piano beautifully and would do so for hours, which was a delightful treat."[14] After Butler's departure, according to her descendants, she "would go from room to room swinging a shovelful of live coals to exorcise any contaminating spirits he may have left behind."[15]

A copy of Darwin's treatise soon found a place on Mesopotamia's shelf. "I always delighted in your *Origin of Species* as soon as I saw it out in New Zealand," Butler wrote to Darwin in 1865, "not as knowing anything whatsoever of natural history, but it enters into so many deeply interesting questions, or rather it suggests so many, that it thoroughly fascinated me."[16] The unencumbered New Zealand wilderness amplified the grand scope of Darwin's ideas. "Residing eighteen miles from the nearest human habitation, and three days' journey on horseback from a bookseller's shop, I became one of Mr. Darwin's many enthusiastic admirers," Butler later recollected, explaining how he came to write a philosophical dialogue upon the *Origin of Species* that was serialized, anonymously, in the Christchurch *Press* starting on December 20, 1862.[17]

"This Dialogue, written by some [one] quite unknown to Mr. Darwin, is remarkable from its spirit and from giving so clear and accurate a view of Mr. D[arwin]'s theory," Darwin himself noted at the time. "It is also remarkable from being published in a colony exactly 12 years old, in which it might have [been] thought only material interests would have been regarded."[18] Butler's dialogue extended across seven issues of the *Press*, including a response titled "Barrel-Organs," ascribed to the bishop of Wellington but more likely planted by Butler himself.

According to this critic, Darwin offered "nothing new, but a réchauffée of the old story that his namesake, Dr. Darwin"—a reference to his grandfather Erasmus Darwin—"served up at the end of the last century," while the deeper implications of the Darwinian theory were nothing more than hymns performed in "a church where the psalms are sung to a barrel-organ, but unfortunately the psalm tunes come in the middle of the set, and the jigs and waltzes have to be played through before the psalm can start."[19]

The series concluded with a manifesto titled "Darwin Among the Machines," appearing under the byline "Cellarius" on June 13, 1863. "We find ourselves almost awestruck at the vast development of the mechanical world, at the gigantic strides with which it has advanced in comparison with the slow progress of the animal and vegetable kingdom," the correspondent warned. "It appears to us that we are ourselves creating our own successors, giving them greater power and supplying by all sorts of ingenious contrivance that self-regulating, self-acting power which will be to them what intellect has been to the human race . . . The machines are gaining ground upon us; day by day we are becoming more subservient to them; more men are daily bound down as slaves to tend them; more men are daily devoting the energies of their whole lives to the development of mechanical life."[20]

Butler's crusade was not *against* Darwin but *beyond* Darwin. "He is gone as far as man can go now, he is an ultra-Darwinian," Edward Chudleigh, one of the hired hands at Mesopotamia, noted in his diary for March 19, 1864.[21] After his return to London in October 1864, Butler produced another commentary, "The Mechanical Creation," published under his own initials in the London *Reasoner* of July 1, 1865. "Those who accept the Darwinian theory will not feel inclined to deny that whatever impulse the animal and vegetable kingdoms have sprung from, has been derived from within the natural influences which operate upon this world, and not from any extra natural source," he argued. "What shall we think then? That the resources of nature are at an end, and that the animal phase is to be the last which life on this globe is to assume? We can see no a priori objection to the gradual development of a mechanical life, though that life shall be so different from ours that it is only by a severe discipline that we can think of it as life at all."[22]

Butler spun these ideas into *Erewhon; or, Over the Range*, casting the antipodean landscape of the Rangitata district as an isolated subalpine valley whose inhabitants proscribe all mechanized technology more advanced than a certain species of mangle, or laundry wringer, to preclude the development of life and intelligence among machines. *Erewhon*, published anonymously in 1872, met with enthusiastic reviews and sold out its first printing in three weeks. "The reviewers did not know but what the book might have been written by a somebody whom it might not turn out well to have cut up, and whom it might turn out very well to have praised," Butler would later explain.[23] Unfortunately, "the demand fell off immediately on the announcement of my being the author,"[24] or, as Butler's friend and biographer Henry Festing Jones put it, "as soon as the *Athenaeum* announced that *Erewhon* was by a nobody the demand fell 90 percent."[25]

In May 1872, Butler wrote to Darwin apologizing "about a portion of the little book *Erewhon* which I have lately published, and which I am afraid has been a good deal misunderstood. I refer to the chapter upon Machines . . . I am sincerely sorry that some of the critics should have thought that I was laughing at your theory, a thing which I never meant to do, and should be shocked at having done."[26] In reply, Darwin invited Butler to visit the Darwin household at Down. Butler stayed with the Darwins for a weekend, a visit, wrote the Darwins' houseguest, "of which I shall always retain a most agreeable recollection."[27]

Darwin complimented Butler on his writings, not only his evolutionary dialogues but also his 1865 pamphlet *The Evidence for the Resurrection of Jesus Christ, as Given by the Four Evangelists, Critically Examined* and *The Fair Haven*, an expansion of these doubts about the resurrection and other weaknesses of Christian orthodoxy, subtitled *A Work in Defence of the Miraculous Element in Our Lord's Ministry upon Earth, Both as Against Rationalistic Impugners and Certain Orthodox Defenders* and attributed to "the late John Pickard Owen," who, like Erewhon, did not otherwise exist.[28] Butler returned the compliment. "*Origin of Species* had, as he told me, completely destroyed his belief in a personal god," John Butler Yeats would later explain. "For him *The Origin of Species* was the book of books."[29]

Despite Darwin's hospitality, Butler became obsessed with finding flaws in Darwin's theory, precipitated by Darwin's failure to credit the contributions of prior evolutionists, including Erasmus Darwin, Georges Buffon, Patrick Matthew, Robert Chambers, and Jean-Baptiste Lamarck. Butler's criticism of Darwin's incomplete acknowledgments escalated into an attack on the foundations of Darwinism itself, which Butler labeled *neo*-Darwinism to distinguish it from the theories of those who had preceded Charles.[30]

"The affair has annoyed and pained me to a silly extent," Darwin confided to Thomas Huxley. "Until quite recently he expressed great friendship for me, and said he had learnt all he knew about evolution from my books."[31] On the advice of Huxley, Darwin withheld comment, although some thirty-six references to "my" theory in the first edition of *Origin of Species* were deleted from subsequent editions of the book. "Has Mivart bitten him and given him Darwinophobia?" asked Huxley in a letter sent to Darwin in 1880. "It's a horrid disease and I would kill any son of a [Huxley inserts a sketch] I found running loose with it without mercy."[32]

Butler's campaign against the scientific establishment gained him few friends. "When Mr. Butler's *Life and Habit* came before us we doubted whether his ambiguously expressed speculations belonged to the region of playful but possibly scientific imagination or unscientific fancies, and we gave him the benefit of the doubt," observed the *Saturday Review* upon the publication of *Evolution, Old and New.* "He has now settled the question against himself."[33]

The theory of evolution, to Butler, did not preclude belief in either a final cause or intelligent design. He favored Erasmus Darwin over Charles. "The world itself might have been generated, rather than created by the Almighty fiat," Erasmus Darwin had argued in 1794. "If we may compare infinities, it would seem to require a greater infinity of power to cause the causes of effects, than to cause the effects themselves."[34] To Butler, the evolution of intelligence and the intelligence of evolution were two sides of the same coin. Species, he explained, "thus very gradually, but nonetheless effectually, design themselves."[35]

"The attempt to eliminate intelligence from among the main agencies of the universe has broken down," he announced. "There is design, or cunning, but it is a cunning not despotically fashioning us from without as a potter fashions his clay, but inhering democratically within the body which is its highest outcome, as life inheres within an animal or plant."[36] Butler believed the bottom-up intelligence of evolution was as much the workings of the mind of God as intelligence delivered top down. He saw each species as the collective embodiment of an intelligence transcending the life of its individual components as surely as we transcend the life and intelligence of our component cells.

"What I wish is, to make the same sort of step in an upward direction which has already been taken in a downward one, and to show reason for thinking that we are only component atoms of a single compound creature, life, which has probably a distinct conception of its own personality though none whatever of ours,"[37] he argued, much as "the component cells of our bodies unite to form our single individuality, of which it is not likely they have a conception, and with which they have probably only the same partial and imperfect sympathy as we, the body corporate, have with them."[38]

Butler concluded that it was impossible to draw a precise distinction between living and nonliving things, or to give a precise definition of life that would not, sooner or later, include machines. "The only thing of which I am sure," he argued in 1880, "is that the distinction between the organic and inorganic is arbitrary; that it is more coherent with our other ideas, and therefore more acceptable, to start with every molecule as a living thing, and then deduce death as the breaking up of an association or corporation, than to start with inanimate molecules and smuggle life into them."[39]

Butler's argument is no longer as radical as it seemed at the time. Self-replicating molecules have molded the world to their own ends. The closer we examine the details of living organisms, the more we see the workings of molecular machines, and the more we observe our most complex technological systems, the more we find them behaving as living things. Intelligence, whether natural or artificial, is an evolutionary system, with internal models of reality evolving by natural selection to find the best fit. Evolutionary systems are displaying emergent intelligence everywhere we look. Horizontal gene transfer and other non-neo-Darwinian evolutionary mechanisms are far more prevalent than we once thought. Transcribing the genomes of living things, we find the exchange of genetic sequences between species is proving to be closer to the elder Samuel Butler's view of languages speciating, yet still exchanging words, than to the neo-Darwinists' germ-line-only genetic tree.

Non-Darwinian evolution, having proliferated among machines, is now extending to living organisms whose genetic code is being stored, replicated, and optimized within digital computers and then redistributed among living cells. Is technology taking control of the living cell, or is life adopting the electronic distribution of its code, as eukaryotic cells adopted self-replicating processes that might have been independent, or parasitic, before they were appropriated by the cell?

Erewhon contained three central messages: machine intelligence will supervene upon human intelligence as surely as our own intelligence supervened upon that of our individual cells; self-reproduction is inevitable once evolution takes hold among machines; and there is no future in trying to turn back the clock. "There is no security . . . against the ultimate development of mechanical consciousness, in the fact of machines possessing little consciousness now," Butler argued, warning of "the descent of conscious (and more than conscious) machines from those which now exist."[40] As to self-reproduction, "surely if a machine is able to reproduce another machine systematically, we may say that it has a reproductive system," he added. "What is a reproductive system, if it be not a system for reproduction? And how few of the machines are there which have not been produced systematically by other machines?"[41]

The harbor of Port Lyttelton, seven miles southeast of Christchurch, was formed by the crater of an extinct volcano, surrounded by steep hills. When Butler arrived in January 1860, communication between Lyttelton and Christchurch was either by a rough bridle path overland or around the exposed headlands via sea. The colonists soon linked the two communities with an electrical telegraph, the first in New Zealand, whose inauguration on July 1, 1862, inspired a letter appearing in the Christchurch *Press* on September 15, 1863. "Why should I write to the newspapers instead of to the machines themselves, why not summon a monster meeting of machines, place the steam engine in the chair, and hold a council of war?" the anonymous "mad correspondent" asked. "I answer, the time is not yet ripe for this . . . Our plan is to turn man's besotted enthusiasm to our own advantage, to make him develop us to the utmost, and find himself enslaved unawares."

"My object is to do my humble share towards pointing out what is the ultimatum, the *ne plus ultra* of perfection in mechanized development," the correspondent continued, "even though that end be so far off that only a Darwinian posterity can arrive at it. I therefore venture to suggest that we declare machinery and the general development of the human race to be well and effectually completed when—when—when—Like the woman in white, I had almost committed myself of my secret. Nay, this is telling too much. I must content myself with disclosing something less than the whole. I will give a great step, but not the last. We will say then that a considerable advance has been made in mechanical development, when all men, in all places, without any loss of time, are cognizant through their senses, of all that they desire to be cognizant of in all other places, at a low rate of charge, so that the back country squatter may hear his wool sold in London and deal with the buyer himself—may sit in his own chair in a back country hut and hear the performance of Israel in Ægypt at Exeter Hall—may taste an ice on the Rakaia, which he is paying for and receiving in the Italian opera house Covent garden. Multiply instance ad libitum—this is the grand annihilation of time and place which we are all striving for, and which in one small part we have been permitted to see actually realised."[42]

To Samuel Butler, the Lyttelton-Christchurch telegraph line contained the germs of the internet and World Wide Web, much as the Erewhonian prohibition against pocket watches presaged the devices that now bind us to a supervening intelligence with every click.

I first read *Erewhon* by the light of a coal-oil lamp in a tree house in British Columbia, a hundred years after its original publication in 1872. Had Samuel Butler taken passage to the one-year-old colony of British Columbia, not New Zealand, in 1859, *Erewhon* might have been set at the head of one of the labyrinthine mainland inlets, or on Vancouver Island's Forbidden Plateau.

Living without electricity or telephone, I succeeded, briefly, in turning back the clock. But there was no escape. While I retreated to the coastal rain forest, my sister Esther joined the inner circle of those pushing the technological advance. Older than me by twenty months, she had left for Radcliffe College at age sixteen, spending much of her time as a student working on the *Crimson*, "the Harvard newspaper that all those famous people once worked on," as she described it.[43] After graduating, she took a job as a fact-checker at *Forbes*. She was soon writing articles as well as checking them, before leaving *Forbes* to work as an analyst for a series of Wall Street investment firms where she became known for the clairvoyance of her reports.

Three thousand miles away, on weekdays, I walked the quarter-mile trail through the Belcarra forest and across the clearing that had once been the Tsleil-Waututh village site, to a cluster of olive-green Canada Post rural mailboxes at the end of the new road that now brought mail to Belcarra by land rather than by boat. Sometimes I opened the padlocked compartment to find one of Esther's dispatches from New York. The earliest I saved a copy of is dated July 19, 1978. It had been commissioned by New Court Securities concerning a venture named Federal Express.

"Using its own fleet of 32 Falcon Jets, six Boeing 727s, and a varying number of leased aircraft, together with 1,000 radio dispatched leased vans . . . about 30,000 packages . . . are flown each evening to a 'hub,'" Esther reported. Her verdict was favorable, and New Court Securities and its clients made a lot of money on the resulting bet. She added a prophetic comment: "And it runs a sophisticated data-processing network, which one day might conceivably transmit customers' messages as well as its own." At the time, the only publicly accessible digital information networks were proprietary services such as Prodigy and CompuServe, or national utilities such as France's Minitel. The chief information officer of Federal Express was James Barksdale, who would go on to become founding CEO of Netscape, launching the commercialization of the open-standards World Wide Web.

Esther compiled her reports on an Apple II computer, one of the first generation of microcomputers to feature an operating system allowing the user to compose natural-language text on a third-party word processing program without having to master more than a handful of commands in the language used by the machine itself. This was the pre-Cambrian era of personal computing, shared by Apple with Amiga, Commodore, Atari, Tandy, and others now long extinct. The new operating systems, and the expanding hierarchy of languages they were co-evolving with, bridged the gap between human language and machine language, enabling humans to more easily operate computers while enabling computers to more easily operate humans. The command line runs both ways.

My interest in the Institute for Advanced Study's Electronic Computer Project that had permeated my childhood was rekindled, and on one of my periodic visits to the University of British Columbia library on Point Grey in Vancouver I borrowed not only a fresh installment of eighteenth-century North Pacific voyages but also John von Neumann's *Design of Computers, Theory of Automata, and Numerical Analysis*, this being volume 5 of his *Collected Works*. One of von Neumann's lectures, "Probabilistic Logics and the Synthesis of Reliable Organisms from Unreliable Components," given in 1952, struck an Erewhonian chord, prompting my own unpublished version of Butler's letter to the Christchurch *Press*.[44]

"The primary language of the nervous system is *statistical* in character," von Neumann had explained. "From this flow of statistical information, several orders of language removed, results our own perception of mathematics and the whole of human consciousness and thought," I now added. "This hierarchy of languages, von Neumann theorized, would characterise complex automata in general. In a machine of a certain critical size, the flow of information, in whatever fundamental form, would exhibit a tendency to convey information in other *more meaningful* forms.

"Machines are seen behaving more and more like human beings, and human beings are seen behaving more and more like machines," I continued. "Human beings have become *statistical* links in this digital electronic chain. A human being today will be counted on—at first digitally and then statistically—to behave, or output, *more or less probably* in a certain mathematically predictable way. The *more or less probably* is no obstacle to the synthesis from these unreliable component parts (human beings) of a reliable, possibly *self-organizing* machine."[45]

In Leibniz's logical universe, all ideas could be formulated, given numerical identity tags, and logically recombined. Leibniz suspected, as Gödel and Turing made explicit, that a mechanical, deterministic system could nonetheless generate creative, unanticipated results. He wanted to see such machines get built. Von Neumann, who did build such a machine, understood that the real world—the territory outside the reservation—was a statistical, probabilistic mess. If you wanted to understand it, let alone control it, better to build probabilistic machines.

In 1982, Esther was hired as managing editor by Benjamin Rosen, an electronic engineer turned investor who produced *The Rosen Electronics Letter* and an associated subscriber conference that was tracking the integrated circuit business as the numbers of chips produced, and the numbers of transistors imprinted on each chip, began to explode. As discrete semiconductors subdivided, like cells in a developing embryo, to become integrated circuits and eventually full-fledged computers on single chips, Rosen's ventures had been swept along by the flood. Microprocessors, enabling inexpensive desktop computer systems, fueled the demand for codes that could run on them. Upstarts that came up with viable codes took the lead over established companies that were encumbered by legacy systems that had to be pushed out of the way first.

Esther observed the new industry "like a good field biologist," as Stewart Brand, the editor of the *Whole Earth Catalog*, put it, wielding a "wicked remote fondness for her obligingly complex subject."[46] Rosen's investments had now spawned conflicts of interest that he resolved by selling the newsletter and conference to Esther in 1983. She renamed the newsletter *RELease 1.0* and rebranded the annual subscriber conference as PC Forum, described by *The Wall Street Journal* as one that "computer field leaders don't dare miss." She shifted the focus to software, and *RELease 1.0* became the chronicle of an industry that was otherwise moving too fast for words. The early PC Forums were like middle school dances, with investment bankers and software developers lined up on opposite sides of the room. Esther spoke the language of the bankers while wearing a T-shirt and jeans.

I attended the four-day conference, as Esther's little brother, for twenty-two years. The other 361 days of the year I was an outsider to the industry, so the changes appeared sudden and discrete. Personal computing was still emerging out of chaos in 1983. How, and to what degree, could the competing operating systems and applications communicate with each other, beginning with physical obstacles such as whether they could even read from and write to each other's 360-kilobyte floppy disks? Memory capacity, processor speed, and disk space grew larger as the machines grew smaller from year to year. Some changes were incremental, and some were sudden leaps, as when floppy disk drives were introduced that could read and write to *both* sides of a single disk. Portable computers started showing up. At first they were the size and weight of sewing machines, required 110-volt power, and could run only one program, loaded on five-inch floppies, at one time.

Desktop computers had multiple expansion slots so that features like graphics, numerical coprocessing, memory cache, audio, and video could be added as they were introduced. One year everyone was adding Ethernet cards and connecting to the internet via a T1 line patched into the hotel. A few years later, by which time most attendees had laptop computers with an external card slot, wireless internet cards were handed out, with a temporary wireless network access point deployed.

Success did not always correspond to being first. There was GO, an early tablet-sized, pen-based computer; Grid, an archetypal ruggedized notebook-form computer; the Apple Newton, a handheld all-purpose computer with a primitive touch screen and lacking only the wireless network by which its eventual successor would conquer the world; the PalmPilot; the Treo (an early "smart" mobile phone); and many, many others, including Metricom, the decentralized packet-switched wireless data network that sought to connect everything together the way we take for granted today. Occasionally, a breakthrough product, like Tetris, or the nascent Facebook, trying to scale beyond the Harvard campus, showed up.

Progress was measured through the adoption of common standards like ASCII character sets, scalable fonts, Ethernet, PostScript, PDF, and internet protocol—or the enforcing of de facto standards like MS-DOS by market dominance when mutual agreement failed. The success of the

Apple II was followed by the Apple Lisa, a market failure, followed in turn by the Apple Macintosh, which proved so successful that Steve Jobs was pushed aside, with John Sculley, from Pepsi, assuming control.

Jobs returned to PC Forum in 1989 to introduce his NeXT computer, whose Unix-based operating system, more than any other progenitor, descended to the iPhone of today. NeXT was a sleek, jet-black cube among a sea of beige Macintoshes, IBM PCs, and PC clones. Over late-night drinks in Jobs's suite at the hotel, half the guests were probing, technically, at the new computer and its unprecedented suite of native programs, and half were poring over the user manuals, produced to the standards of high-end European art books and spread out on one of the king-sized beds.

A direct successor to the machine that Julian Bigelow's handful of engineers had assembled from 3,474 vacuum tubes in 1951 was now running at twenty-five million instruction cycles per second in a single silicon chip within a twelve-inch magnesium cube on the coffee table in Steve Jobs's suite. The NeXT cube sat there like the obelisk in Stanley Kubrick's *2001*, waiting for a human touch to awake. Yet its architectural efficiency had not changed in thirty-three years. At any given moment most of its logical elements were sitting idle, waiting to be given the next instruction and memory address.

Bigelow never intended this inefficient architecture, invoked as an emergency solution to von Neumann's hydrogen bomb problem in 1946, to be perpetuated ever since. The logical components of a modern digital computer, he observed, despite being capable of operating continuously at high speed, "are interconnected in such a way that on the average almost all of them are waiting for one (or a very few of their number) to act. The average duty cycle of each cell is scandalously low."[47] To compensate for these inefficiencies, processors execute billions of instructions per second. How can programmers supply enough instructions—and addresses—to keep up?

Iterative, recursive processes are the only way that humans are able to generate instructions fast enough. "Electronic computers follow in-structions very rapidly, so that they 'eat up' instructions very rapidly, and therefore some way must be found of forming batches of instructions very efficiently, and of 'tagging' them efficiently, so that the computer is kept ef-fectively busier than the programmer," Bigelow observed. "This may seem like a highly whimsical way of characterizing a logically deep question of how to express computations to machines. However, it is believed to be not far from an important central truth, that highly recursive, conditional and repetitive routines are used because they are notationally efficient (but not necessarily unique) as descriptions of underlying processes."[48]

The personal computer, with its user-friendly operating systems, gave people with little or no experience in programming access to computers, while making an unlimited supply of instructions available to the ma-chines. At first, the machines did one thing at a time. Then codes learned to "terminate and stay resident," going into suspended animation but staying resident in memory, ready to be summoned on command. Soon, operating systems evolved that could run multiple processes at once. The keyboard was supplemented by a growing menagerie of input devices, beginning with the light pen, which evolved into the mouse, trackball, track pad, and a host of variants, including the touch screen, now the dominant interface between the two worlds.

When the IAS engineers began their physical realization of Alan Tur-ing's universal machine, the chief technical obstacle was the memory: How to construct a digital storage matrix that would be randomly accessible at the speed of light? Their improvised solution was to project a 32-by-32 array of charged spots on the glass face of a standard off-the-shelf five-inch cathode-ray oscilloscope tube, forming a matrix of 1,024 individual capacitors that could hold an electrostatic charge long enough to remem-ber the state of each spot between refresh cycles traced by the electron beam. They designed control circuits that could read or write the state of any selected individual memory location, with twenty-four-microsecond access time, during the intervals between refresh cycles, the discrimina-tion between a zero and a one requiring amplifying the faint signal thirty thousand times. The resulting memory organ, known as a Williams tube after F. C. Williams of Manchester, U.K., who, with Tom Kilburn, first developed a serial-access implementation of it, was, as Bigelow noted, "one of mankind's most sensitive detectors of electromagnetic environmental disturbances."[49]

"We learned, somewhat to our sorrow," recalls James Pomerene, the lead memory engineer, "that it remembers! Big surprise! But among other things, it remembers any noise that ever happened. Right? So it sits there and you're hoping it remembers a one, and some noise comes along and it would turn from one to a zero, and it stays a zero because it's remembering now a zero. So a memory turned out to be a very effective observer of noise."[50] The glass faces of the tubes had to be shielded from any contact with the outside world. The luminous, charged spots could be observed dancing on the screen, but the slightest touch and computation would grind to a halt.

The capacitative touch screen, a living relative of the extinct Williams tube, takes advantage of this susceptibility to local disturbances to translate the touch of a human fingertip, one of our most sensitive channels of communication, directly into machine-readable bits.

Samuel Butler's *Erewhon* is inhabited by the survivors of a civil war between the Machinists and the Anti-machinists. The Anti-machinists, having allowed an exception for advanced weapons, emerge victorious, and the evolution of technology is brought to a halt. Toward the end of his life, Butler returned to *Erewhon*, but his *Erewhon Revisited*, published in 1901, was concerned more with skewering the contradictions of religious faith and family relationships than addressing the further evolution of machines.

Among Butler's notebooks are fragments for an *Erewhon Revisited* where he imagined the Erewhonians giving up their resistance to progress and coming around to the side of the machines. "Let automata increase in variety and ingenuity till at last they present so many of the phenomena of life that the religious world declares they were designed and created by God as an independent species," he speculated. "The scientific world, on the other hand, denies that there is any design in connection with them, and holds that if any slight variation happened to arise by which a fortuitous combination of atoms occurred which was more suitable for advertising purposes (the automata were chiefly used for advertising) it was seized upon and preserved by natural selection."[51]

Automata, advertising, and natural selection are an explosive mix. Google's introduction of AdWords, monetizing not just language, already coded, but also *meaning*, not fully coded yet, was the equivalent of Lee de Forest's introducing the control grid into Fleming's vacuum tube. Internet advertising drives a global high-gain amplifier connecting the reward sought by computers (more machine cycles and instructions) to the reward sought by humans (more of the stimulation now returned with every click). We set loose an evolutionary system that rewards machines that learn to control both how we feel and what we think.

We are now Erewhonians, and the war is over, except the Machinists, not the Anti-machinists, carried the field. Perhaps the best we can do, as Butler suggested, is embrace the rise of advertising-driven intelligence and follow wherever it leads. "Why may not there arise some new phase of mind which shall be as different from all present known phases, as the mind of animals is from that of vegetables?" Butler asked. "When we reflect upon the manifold phases of life and consciousness which have been evolved already, it would be rash to say that no others can be developed, and that animal life is the end of all things. There was a time when fire was the end of all things."[52]

There is no escaping the machines. "Man's very soul is due to the machines; it is a machine-made thing: he thinks as he thinks, and feels as he feels, through the work that machines have wrought upon him," Butler argued. "This is the art of the machines—they serve that they may rule."[53] The machines we call "servers" have become our masters, while we are becoming serfs.

This future need not be as bleak as it appears. "There is reason to hope that the machines will use us kindly, for their existence will be in a great measure dependent upon ours; they will rule us with a rod of iron, but they will not eat us; they will not only require our services in the reproduction and education of their young, but also in waiting upon them as servants; in gathering food for them, and feeding them; in restoring them to health when they are sick," Samuel Butler argued. For those who care for the machines, life will be good. "There is no occasion for anxiety about the future happiness of man so long as he continues to be in any way profitable to the machines; he may become the inferior race, but he will be infinitely better off than he is now."[54]

How is "better off" defined? What value function defines a win? "The game that nature seems to be playing is difficult to formulate," Stanislaw Ulam observed in 1966. "When different species compete, one knows how to define a loss: when one species dies out altogether, it loses, obviously. The defining win, however, is much more difficult because many coexist and will presumably for an infinite time; and yet the humans in some sense consider themselves far ahead of the chicken, which will also be allowed to go on to infinity."[55]

I saw a baby playing with an iPad," my daughter, Lauren, reported from San Francisco International Airport in 2012, "so young it was still drinking mother's milk."

"May not man himself become a sort of parasite upon the machines?" Samuel Butler had asked in *Erewhon*'s "Book of the Machines." "An affectionate machine-tickling aphid?"[56]

8.

NO TIME IS THERE

Five hundred and seventy-five miles north of Fort Bowie, Arizona, the headquarters of General Nelson Miles's heliographic intelligence network during the closing chapter of the Apache campaign, is a million-square-foot complex operated by the National Security Agency. The desert air is no longer as transparent as it was in 1886. Where the U.S. Army Signal Corps gathered hand-modulated beams of sunlight relayed in the open from one mountaintop to the next, the National Cybersecurity Initiative Data Center, at Camp Williams, Utah, now gathers signals transmitted through fiber-optic cables as otherwise imperceptible pulses of light.

The Utah facility can store more than ten trillion times the traffic conveyed by the heliograph network during the entire Apache campaign. It draws sixty-five megawatts of electric power, consumes twelve hundred gallons of cooling water per minute, and cost two billion dollars by the time it became operational in 2014. The precise capabilities of the repository are unknown. All forms of communication, along with associated metadata, such as who communicated with whom and when, are stored in machine-readable form.

Most digital telecommunications are stored in a series of temporary buffers along the way, a practice that descended from nineteenth-century telegraph networks to early twentieth-century data network exchanges, where incoming data streams were buffered to punched paper tape before being routed, sometimes by clerks on roller skates, to the designated outgoing line. We accept this intermediate storage, replication, and relaying of our communications, assuming the buffers to be volatile over short periods of time. Once a buffered message was relayed to the next node in the network, the tape was thrown away. The Utah facility is a centralized nonvolatile buffer that persists indefinitely, as if all the paper tape ever punched were saved in searchable form.

Until a given communication is linked to an authorized surveillance target, it is not supposed to be read. But in the age of machine intelligence, what is meant by "read"? The presumption that a communication, encoded as a number, is "unread" until a human being views it or listens to it no longer holds. The moment a number enters the working memory of a digital computer, it is read. Are certain forms of arithmetic acceptable and others out of bounds?

There are two approaches to machine intelligence: assembling fragments of information into a framework, like a kayak; or discarding information until knowledge is revealed, like a dugout canoe.

A digital universe is populated by two species of bits: differences that are varying in time but invariant in space, and differences that are varying in space but invariant in time. Bits can be stored over time as memory, or communicated across distance as code. Digital computers translate between these two forms of information—structure and sequence—according to definite rules. These powers of translation are more general than the arithmetical functions for which they were first invoked. Nature, too, discovered a method for translating sequences (of nucleotides) into structures (of proteins)—and back. Once this loop is established, evolution will do the rest.

Digital computing goes back to Leibniz, who credited the discovery of the underlying binary arithmetic to the Chinese. "The 64 hexagrams [of the *I Ching*] represent a binary Arithmetic . . . which I have rediscovered some thousands of years later," he reported to Nicolas de Remond in his *Discourse on the Natural Theology of the Chinese*. "In Binary Arithmetic there are only two signs, 0 and 1, with which we can write all numbers . . . I have since found that it further expresses the logic of dichotomies which is of the greatest use."[1]

Leibniz constructed a wheel-based mechanical calculator in 1672, inspired by a predecessor built by Blaise Pascal. He then imagined a digital computer that would operate without wheels using binary arithmetic, as he described in 1679. "This [binary] computation can be executed by a machine without wheels, in the following manner," he explained. "A container shall be provided with holes in such a way that they can be opened and closed. They are to be open at those places that correspond to a 1 and remain closed at those that correspond to a 0. Through the opened gates small cubes or marbles are to fall into tracks, through the others nothing. It [the gate array] is to be shifted from column to column as required."[2]

Leibniz had invented the shift register: a fundamental engine of the digital age. A shift register stores a small number of bits in a linear array and, on command, shifts them by one place either right or left—as if you have a row of eggs in an egg carton and were able to take them all out and put them all back, shifted by one position, without ever breaking or misplacing an egg.

Shifting a string of binary digits one place to the left multiplies the corresponding number by two; shifting it to the right divides it in half. This operation, logically independent of the mechanical gates, electromechanical relays, electronic vacuum tubes, or solid state transistors by which it is implemented, remains etched into microprocessors at submicron scale. Leibniz did not live to see his invention realized, just as he did not live to see his *Monadology* brought to life by the discovery of elementary particles, entangled by quantum mechanics into a system reminiscent of his universe of monads, or "little minds." He died 268 years before his shift register was reinvented, and 25 years before the survivors of the Great Northern Expedition returned from America with the answers to the questions he had posed to Peter the Great at Bad Pyrmont shortly before his death.

Five-stage gas-filled thyratron shift registers were used in the Colossus code-breaking computers built under the direction of Thomas Flowers in England during World War II, and vacuum-tube shift registers were built experimentally, also during the war, under Jan Rajchman's supervision at RCA. After the war, it was the small team of electronic engineers working under Julian Bigelow, Herman Goldstine, and John von Neumann at the Institute for Advanced Study in Princeton, New Jersey, who brought Leibniz's vision to life: with a voltage gradient serving in place of gravity, pulses of electrons serving in place of marbles, vacuum tubes serving in place of mechanical gates, and wires serving in place of tracks.

Von Neumann and his government sponsors, under pressure to evaluate the radiation hydrodynamics underlying the feasibility of the hydrogen bomb, required a machine that could operate on forty-bit numbers in parallel at kilocycle speeds. Forty-stage shift registers had to be coaxed into operation with a reliability far beyond anything achieved so far by electronic engineers. "It was easy to build a two-stage register that worked reliably," says Jack Rosenberg, the fourth engineer to join the group. "When a third stage was added occasional errors crept in. Adding a fourth stage made the register useless. We discovered that the electrical characteristics of the vacuum tubes were very different from the specifications published for them in tube handbooks, even when the tubes were new." Questioned about the discrepancies, the manufacturers answered that "no one else had ever complained about their product, and they had enough customers without us."[3]

The engineers switched from designing circuits according to the published tube specifications to what Bigelow, taking a page from the fashion industry, termed "the new look." As Ralph Slutz, an engineer who went on to build the SEAC for the National Bureau of Standards, the first working version of the IAS design, explained, "We tested a batch of a thousand tubes and took the weakest tube we found and the strongest tube we found, and then allowed an extra 50% safety factor over that."[4] In contrast to mechanical systems that function more reliably at slower speed, Bigelow understood that "increasing speed may actually increase certainty rather than the reverse," because vacuum tubes are weakened by age, not use, and "suffer accidental failures in proportion to their population," not their operating speed.[5]

Each bit was stored in a circuit commonly known as a flip-flop, but more accurately termed a "toggle," implemented within a mass-market twin-triode miniature vacuum tube: the 6J6. The lightbulb-sized vacuum tubes of Fleming and de Forest had been reduced in size to a little smaller than a wine cork, with two complete triodes, sharing a common heater element, contained within a single glass envelope having seven connector pins. The pair of triodes were connected so that one or the other, but not both, was in an energized, conducting state, the alternatives signifying either a zero or a one. To transfer data between adjacent toggles, the state of each individual toggle was replicated upward into a temporary register, the lower register was cleared, and then and only then were the data shifted, diagonally, back down into the original register. Electrons were always looking for an escape.

"Information was first locked in the sending toggle; then gating made it common to both sender and receiver, and then when securely in both, the sender could be cleared," Bigelow explained. "Information was never 'volatile' in transit; it was as secure as an acrophobic inchworm on the crest of a sequoia." Data were handled the way ships are moved through locks in a canal. "We did not move information from one place to another except in a positive way," emphasizes James Pomerene.[6]

The edict that "absence of a signal should never be used as a signal," a principle that Bigelow had arrived at in his work on anti-aircraft fire control with Norbert Wiener in World War II, drove the design of the computer his team now built.[7] The reliable communication of information between thousands of memory locations, thousands of times per second, was the chief obstacle to getting the computer to work. Von Neumann, Bigelow, and colleagues including Stan Ulam were already thinking about what might develop in a much larger, or even unbounded, matrix of communicating cells. "We enjoyed some interesting speculative discussions with von Neumann at this time about information propagation and switching among hypothetical arrays of cells," adds Bigelow, "and I believe that some germs of his later cellular automata studies may have originated here."[8]

In June 1947 a prototype ten-stage shift register was completed. During July and August it was "tested by causing a fixed pattern of 1's and 0's to be shifted around more or less continuously. The pattern was shifted by a group of 11 pulses, each less than 1/2 microsecond wide, spaced 4 microseconds apart, each group being spaced 1 millisecond apart."[9] Neon indicator lamps displayed the state of each individual toggle, and if the shift register was functioning correctly, the test pattern would appear stationary to the human eye.

After six more design iterations, a production model was settled upon, and by the end of 1948 three shift registers, each with forty stages, had been "interconnected in two different arrangements to form closed loops of 120 binary digits and shifted a place at a time around the loop," at a rate of three microseconds per shift. The test was run seventy-five hours on left shifts and about twenty-five hours on right shifts, for a total of a hundred billion shifts.[10]

The completed shift registers were incorporated into a high-speed electronic stored-program digital computer, christened the Mathematical and Numerical Integrator and Computer (MANIAC), that became operational in 1951. The forty-bit strings of code that populated its registers and memory organs were termed "words" and could be numbers, instructions, address references, encoded natural language, primitive numerical organisms, or anything else that could be expressed in binary form.

This mingling of data with instructions broke the distinction between numbers that *mean* things and numbers that *do* things, and all hell broke loose, starting with the hydrogen bomb, as a result. Until stored-program digital computers, numbers represented things. Now coded instructions, termed "order codes," were given the power to do things—including the power to invoke another instruction or make copies of themselves. There was a vocabulary of some twenty-nine order codes, conveyed in hexadecimal notation by the human operator, with none of the intervening hierarchy of interpretive languages we take for granted today.

Strings of bits gained the power of self-replication, just like strings of DNA. Thus began a chain reaction, with the order codes persisting largely unchanged, like the primordial alphabet of amino acids, over the seventy years since they were first released.

The MANIAC's descendants, replicated first in vacuum tubes, next in discrete semiconductors, and now in monolithic silicon, are characterized by word length, governing how much memory they can address, and clock speed, governing how many instructions they can execute in a given period of time. The benchmarks have advanced to sixty-four-bit words and billions of instruction cycles per second. The underlying "clocks," however, are there not to measure time but to serve as a clockwork escapement regulating an orderly sequence of events. In the digital universe, time as we know it does not exist.

In the analog universe, time is a continuum. Any two moments, no matter how close, have other moments in between. In the digital universe, there is no continuum, only a finite if unbounded series of discrete steps. By convention, we use a sequence of regular pulses provided by a "clock" to count these steps, but it is the number of increments, not the passage of time, that counts.

"The MANIAC machine, it didn't have anything like a pulser in it—no clocks, no pulsers, no nothing," Bigelow explains. "It was all of it a large system of on-and-off, binary gates. No clocks. You don't need clocks. You only need counters. There's a difference between a counter and a clock. Time is not the variable you keep track of. The sequence is what you keep track of. And that's enormously different from a clock. A clock keeps track of time—and a modern general purpose computer keeps track of events. Sequence is different from time. No time is there."[11]

When a string of code is harvested by the Utah data repository, the sequence is preserved, but the constraints of time are removed. Shift registers that are logically no different from those that Leibniz imagined in 1679 translate the incoming streams of time-variant sequences, frame by frame, into memory registers where they become data structures invariant in time. The preserved sequence can be recalled, as needed, to reflect the illusion of time back to us, as if the digital universe were a model of our familiar existence, rather than an alien construct growing in our midst.

To observers in our universe, the digital universe, regulated by ever-faster "clock cycles," appears to be speeding up, doing more and more in a given interval of continuous time. To observers in the digital universe, our universe, doing less and less over a given number of increments, appears to be slowing down.

At the end of the seventeenth century, Leibniz and Newton both claimed priority for the invention of the calculus, while their contemporary Robert Hooke, curator of experiments to the Royal Society of London, nursed a private grudge over his own theory of gravitation and celestial mechanics, of which, said John Aubrey, "Mr. Newton haz made a demonstration, not at all owning he receiv'd the first Intimation of it from Mr. Hooke."[12]

When Leibniz exhibited his calculating machine to the Royal Society in January 1673, Hooke complained that he "could not perceive it ever to be of any great use" and announced, "I have an instrument now making, which will perform the same effects [and] will not have a tenth part of the number of parts, and not take up a twentieth part of the room."[13] On March 5, 1673, Hooke demonstrated his own arithmetical engine "and showed the manner of its operation, which was applauded," but the invention then disappeared. His *philosophical algebra* claimed to be a more powerful system than Leibniz's *calculus ratiocinator*, but he kept the details to himself. Fearing expropriation of his work, Hooke revealed many of his later inventions only in the form of cryptic anagrams, carrying the details with him to his grave. "I wish he had writt plainer, and afforded a little more paper," Aubrey complained.[14]

Hooke was less interested in artificial intelligence and more interested, as he put it, in discovering "how the Organs made use of by the Mind in its Operation may be Mechanically understood," not to undermine the existence of the soul or the nature of free will, but to illuminate how the soul does its work. "The Soul, or First Principle of Life, tho' it be an Incorporeal Being," he observed, "yet in performing its Actions, makes use of Corporeal Organs, and without them cannot effect what it wills."[15]

Hooke, who had advanced the design of chronometers with his escapement based on a coiled spring, proposed a "hypothetical explication of memory" based on a coiled series of ideas. "There is as it were a continued Chain of Ideas coyled up in the Repository of the Brain, the first end of which is farthest removed from the Center or Seat of the Soul where the Ideas are formed, which is always the Moment present when considered," he suggested. "And therefore according as there are a greater number of these Ideas between the present Sensation or Thought in the Center, and any other, the more is the Soul apprehensive of the Time interposed."[16]

He estimated the number of thoughts that could be registered per second, hour, day, year, and lifetime—"to take a round sum but 21 hundred Millions"—and then reduced to one hundred million the number that the average person might remember, who "consequently must have as many distinct Ideas." The author of *Micrographia*, an atlas of nature viewed at microscopic scale, drew on his firsthand observations of microorganisms to argue that this many ideas might easily fit inside the brain: "I see no Reason why all these may not actually be contained within the Sphere of Activity of the Soul . . . for if we consider in how small a bulk of Body there may be as many distinct living creatures as are here supposed Ideas . . . we shall not need to fear any Impossibility to find out room in the Brain."[17]

Hooke, remembered for Hooke's law relating stress to strain in elastic bodies, believed there to be a similar scaling law relating the size of creatures to their perception of time. "The sensible Moments of Creatures are somewhat proportion'd to their Bulk," he observed, conjecturing "that the less a Creature is, the shorter are its sensible Moments; and that a Creature that is a hundred times less than a Man, may distinguish a hundred Moments in the time that a Man distinguishes one." He concluded that "many of those Creatures that seem to be very short lived in respect to Man, may yet rationally enough be supposed to have lived, and been sensible of and distinguished as many Moments of time."[18] Technology has extended Hooke's scale of sensible moments in both directions, from the nanosecond moments of a microprocessor to the nonvolatile memory of the Utah data repository and its infinitely coiled springs.

In 1679, the year that Leibniz envisioned the shift register, there appeared in London a new edition of an account of an early project in artificial intelligence, this being a thirteenth-century attempt to build a speaking brass head. *The Famous History of Frier Bacon, Containing the Wonderful Things That He Did in His Life; Also the Manner of His Death, with the Lives and Deaths of the Two Conjurers Bungey and Vandermast* was subtitled *Very Pleasant and Delightful to Be Read.*

Roger Bacon, the principal investigator, was a scholar and magician who ventured beyond astrology and alchemy to embrace sciences far ahead of his time. Said to have been imprisoned for fifteen years by his own Franciscan brothers for the novelty of his ideas, he became known as Doctor Mirabilis and was reputed to have undertaken to defend England against invasion by constructing a wall of brass about the entire island, a project for which he sought to enlist the intelligence of a brazen head.

"To this purpose he got one Frier Bungey to assist him," it was reported, "who was a great scholar and a Magician, (but not to compare to Frier Bacon) these two with great study and pains so framed a head of Brass, that in the inward parts thereof there was all things like as in a natural man's head: this being done, they were as far from perfection of the work as they were before, for they knew not how to give those parts that they had made, motion, without which it was impossible that it should speak: many books they read, but yet could not find out any hope of what they sought, that at the last they concluded to raise a spirit, and to know of him that which they could not attain to by their own studies."[19]

This uncooperative "Devil" was coerced into disclosing the secret formula that would enable the brass head to speak but refused to specify the length of time required for it to take effect. "If they heard it not before it had done speaking," they were warned, "all their labour should be lost." Bacon and Bungey followed the devil's instructions and waited three weeks, with no results. Then Bacon assigned his servant Miles to keep a close watch on the brass head so the two inventors could take a nap.

Miles stood watch while his master slept, and then "at last, after some noise the head spake these two words, *Time Is*. Miles hearing it to speak no more, thought his Master would be angry if he waked him for that, and therefore he let them both sleep, and began to mock the head in this manner: Thou Brazen-faced head, hath my Master took all this pains about thee, and now dost thou requite him with two words, *Time Is*: had he watched with a lawyer as long as he hath watched with thee, he would have given him more, and better words then thou hast yet, if thou canst speak no wiser, they shall sleep till dooms day for me."[20]

Miles kept mocking the brass head: "Do you tell us Copper-nose, when *Time Is*? I hope we Scholars know our times, when to drink, when to kiss our hostess, when to go on her score, and when to pay it, that time comes seldom." After half an hour of this, "the head did speak again, two words, which were these: *Time Was*." Miles still would not wake his master, saying, "If you speak no wiser no Master shall be waked of me," and acted like a fool for another half an hour. Then, without further warning, "this Brazen Head spake again these words; *Time is Past*: and therewith fell down, and presently followed a terrible noise, with strange flashes of fire, so that Miles was half dead with fear: at this noise the two Friers awaked, and wondered to see the whole room so full of smoke, but that being vanished they might perceive the brazen head broken and lying on the ground."[21]

The specter of a bargain for secret knowledge leading to unintended consequences has repeated itself across the history of artificial intelligence, from the legend of the golem to Mary Shelley's *Frankenstein* to John von Neumann's contract with the government that gained him the computing power he so coveted in exchange for the design of the hydrogen bomb. Deep learning, promising the powers of artificial intelligence while refusing to reveal how they work, is the latest secret key.

Leibniz's digital utopia, like all utopias, arrived with flaws. The same numerical bureaucracy that brought us the address matrix enabling the internet also brought the system that forced reservation-bound Apaches, in the 1870s, to wear numbered metal identity tags around their necks.

In November 1871, General Philip H. Sheridan, acting under General Sherman's instructions in reaction to a series of outbreaks in Arizona and New Mexico, ordered that "all roving bands of Indians, for which reservations have been set apart by the Indian commission, under the authority of the President of the United States, will be required to go at once upon their reservations, and not to leave them again upon any pretext whatever."

"So long as they remain upon their reservations in due subordination to the Government, they will be fully protected and provided for; otherwise they will be regarded as hostile, and punished accordingly," Sheridan added, advising General Crook, assigned to execute these orders, that he "may feel assured that whatever measures of severity he may adopt to reduce these Apaches to a peaceful and subordinate condition will be approved by the War Department and by the President."[22]

To distinguish those to be declared hostile, the secretary of war and the secretary of the interior agreed to establish a register of nonhostile Apaches, in which "all male Indians (old enough to go upon the war-path) will be enrolled, and their names will be recorded in a book kept for that purpose, with a full and accurate descriptive list of each person." The notion that "each Indian will be furnished with a copy of his descriptive list, and will be required to carry it always with him," did not survive implementation in the field, with Crook's system of numbered identity tags instituted instead.[23] "So long as any bad Indians remained out in the mountains," John Bourke reported, "the reservation Indians should wear tags attached to the neck, or in some other conspicuous place, upon which tags should be inscribed their number, letter of band, and other means of identification."[24]

The tags helped to regulate the distribution of rations and standardize the confusing-to-the-government system of Apache names, with a regular census of the tagged population making it possible to identify any individuals who had made their escape. "The presence on the reservation of every male adult will be verified once a day, or oftener," General John Schofield ordered. Any Apache without a valid tag was declared hostile and treated accordingly. If an adult male was found, upon census, to be absent, "all his family will be arrested and kept in close custody until he has been captured and punished according to his deserts."[25]

In November 1872, Crook issued a deadline after which no adult males would be allowed to return to the reservation except as prisoners of war, and then launched a relentless pursuit of those who remained at large. On April 9, 1873, after some five hundred fugitives had been killed at the hands of three separate detachments of soldiers and scouts who had each covered some twelve hundred miles in the campaign, he claimed to have "finally closed an Indian war that has been waged since the days of Cortez."[26] There were thirteen more years of intermittent skirmishes and outbreaks until the 410 remaining Chiricahuas were driven on a forced march to the railroad siding at Holbrook and loaded on board the Florida-bound train.

General Miles's heliograph network anticipated the advent of optical data networking, and General Crook's numbered identity tags anticipated the digital identities we are captive to today. The National Cybersecurity Initiative Data Center is only one node in a growing archipelago of data centers mapping communications to these identities over time. Voice recognition, facial recognition, and even the bar-coded signature of DNA are being linked to the universal identity tag of the mobile phone. Anyone without a tag stands out. Even Kurt Gödel, who proved that it was possible to escape the bounds of any formal system, no matter how powerful, by constructing statements that are true yet cannot be proved, would have trouble escaping from a system where not being listed adds you to a special list. The result is a reservation without physical boundaries where time erases no tracks.

In 2018, in Mesa, Arizona, at the foot of the former trail leading to Fort Apache and Apache Pass, Apple Computer opened its global data command center, a two-billion-dollar, 1.3-million-square-foot facility, overseeing the network of data centers that form the digital backplane underlying all the iPhones in the world. The Arizona facility is powered by its own fifty-megawatt solar collector farm. Heliography is back.

Leibniz had envisioned a formal system capturing a binary encoding of all human communication and thought. He failed to interest Peter the Great, but the tsars are now signing on. In China, where binary arithmetic had a five-thousand-year head start, 80 percent of the population have mobile phones. The State Council of the People's Republic of China recently announced "the formation of a credit investigation system that covers the whole society . . . establishing mechanisms to encourage trustworthiness and punish untrustworthiness," through the deployment of an all-encompassing surveillance network linking biometric identity tags, machine intelligence, and indelible individual social credit scores, with the stated goal of "making it so that the trustworthy benefit at every turn and the untrustworthy cannot move an inch."[27]

Seventy years ago, John von Neumann delivered the wonders of digital computing in exchange for the terrors of the hydrogen bomb. "The factor 4 is a gift of God (or of the other party)," he wrote to Edward Teller in 1946 while performing the initial calculations to determine whether a weapon one thousand times more powerful than those that had just destroyed Hiroshima and Nagasaki could be built.[28] We appear to have done well by the bargain so far: gaining a world transformed by digital computing yet unscathed by thermonuclear war.

Unless the other party knew, from the beginning, that digital computers would become the more powerful instrument, in time. Between the no-fly list and the train to Florida is only a series of steps.

9

CONTINUUM HYPOTHESIS

During the years 1865 and 1866 the great plains remained almost in a state of nature, being the pasture-fields of about ten million buffalo, deer, elk, and antelope, and were in full possession of the Sioux, Cheyennes, Arapahoes, and Kiowas, a race of bold Indians, who saw plainly that the construction of two parallel railroads right through their country would prove destructive to the game on which they subsisted, and consequently fatal to themselves," explained General William Tecumseh Sherman, who, as commander of the U.S. Army in the aftermath of the Civil War, made it his business to ensure that "the fate of the buffalo and Indian was settled for all time to come."[1]

Sherman, the sixth of ten children, was born in Ohio in 1820 at a time when "the Indians still occupied the greater part of the state." His father, a lawyer who spent much of his time on horseback, "seem[ed] to have caught a fancy for the great chief of the Shawnees, Tecumseh," or "shooting star," who was killed in battle, fighting alongside the British against the Americans in the War of 1812.[2] Tecumseh, assisted by his brother Tenskwatawa,

View *of* Kayes Island *the S. End bearing W.S.W. 8 or 9 lea.*

a self-styled prophet described by Thomas Jefferson as "vainly endeavoring to lead back his brethren to the fancied beatitudes of their golden age," had assembled a confederation of tribes resolved to cede no more land to the United States.[3] Their project failed, while Sherman devoted much of his career to making sure it would never have another chance.

Upon graduation from West Point in 1840, Sherman was commissioned as a second lieutenant and assigned to Fort Pierce, Florida, "a spot on earth where fish, oysters, and green turtles so abound . . . that the soldiers regarded it as an imposition when compelled to eat green turtle steaks." By this time, he reported, "the Indians in the Peninsula of Florida were scattered, and the war consisted in hunting up and securing the small fragments, to be sent to join the others of their tribe of Seminoles already established in the Indian Territory west of Arkansas."[4]

As westward expansion encroached upon the already displaced tribes, Sherman advocated for confinement to reservations and against continued recognition of Native Americans as independent nations under the law. He defended the army's 1870 attack on a Piegan village in Montana in which fifty-three women and children were killed, authorized General Crook's pursuit of the Apaches into Mexico, endorsed severe retaliation against the Modocs who led an uprising in California, and argued for the abrogation of the 1868 treaty with the Sioux. He never voiced second thoughts. The construction of the transcontinental railroads and consequent extermination of the buffalo "will help to bring the Indian problem to a final solution," he predicted in 1872.[5]

Sherman made a name for himself, during the Civil War, by marching through Georgia to the sea, crippling the rebel forces by laying waste to their farms and stores of food. He saw the march of civilization toward the Pacific in similar terms. "The close of the civil war," he observed, left "nearly a million of strong, vigorous men who had imbibed the somewhat erratic habits of the soldier; these were of every profession and trade in life, who, on regaining their homes, found their places occupied by others, that their friends and neighbors were different, and that they themselves had changed. These men flocked to the plains, and were rather stimulated than retarded by the danger of an Indian war." Within a generation, the settlers had succeeded, with the army's help, in having "replaced the wild buffaloes by more numerous herds of tame cattle . . . by substituting for the useless Indians the intelligent owners of productive farms."[6] Resistance among the Sioux, Apaches, and other tribes continued, "but they have been the dying struggles of a singular race of brave men fighting against destiny, each less and less violent, till now the wild game is gone, the whites too numerous and powerful; so that the Indian question has become one of sentiment and charity, but not of war."[7]

The North American bison survived only in small groups scattered across remnants of its former range. "During the first three months of the year 1886 it was ascertained . . . that the extermination of the American bison had made such alarming progress . . . that the destruction of the large herds, both North and South, was already an accomplished fact," William Temple Hornaday, chief taxidermist for the U.S. National Museum, reported, estimating that "less than three hundred" remained in the wild in the entire United States.[8]

With only one specimen in the museum's collection, the secretary of the Smithsonian Institution, Spencer F. Baird, "resolved to collect at all hazards, in case buffalo could be found, between eighty and one hundred specimens of various kinds, of which from twenty to thirty should be skins, an equal number should be complete skeletons, and of skulls at least fifty."[9] In the late spring and fall of 1886, Hornaday led two expeditions to the wildlands of Montana between the Yellowstone and the Missouri Rivers, where he collected eleven bulls, eleven cows, and three calves, including one calf taken alive and sent to Washington, where it survived in captivity until July 26, 1886. By the end of 1888, Hornaday estimated the wild bison population was down to eighty-five in the United States.[10]

The open frontier was at an end.

The people who had hunted bison across the open plains were the direct descendants of those who had hunted sea mammals along the shores of

Beringia as the Pleistocene came to a close. As a culture that had flourished in North America for fifteen thousand years was extinguished, there arose a movement, known as the Ghost Dance religion, that was not so much a religion as a refusal to give up hope.

The movement manifested itself across western North America along two main branches, one arising with the prophet Wodziwob, a Paiute from Walker Lake, Nevada, in 1870 and the other arising with the prophet Wovoka, also a Paiute and known by the name Jack Wilson, in 1890. The Ghost Dance movement is remembered in messianic terms: first because it was described by Christians who saw in it a reflection of their own faith; second because its own apostles incorporated elements of Christian belief; and finally because a belief in resurrection is as fundamental as the human awareness of death.

Peter Schonchin, whose father was one of those hanged after the Modoc uprising of 1872, remembered how in 1870 Wodziwob had predicted that "the dead would come from the east when the grass was about 8 inches high . . . The deer and the animals were all coming back . . . The white people would die out and only Indians would be on earth. The whites were to burn up and disappear without even leaving ashes."[11]

Twenty years later, the vision that came to Wovoka, a disciple once removed from Wodziwob, spread more widely, ranging eastward, and culminated in tragedy among the Sioux, who, as James McLaughlin, of Standing Rock Agency, reported in October 1890, "were greatly excited over the near approach of a predicted Indian millennium or 'return of the ghosts,' when the white man would be annihilated and the Indian again supreme, and which the medicine men had promised was to occur as soon as the grass was green in the spring."

In McLaughlin's account, the spirits of the dead warriors "were already on their way to reinhabit the earth, which had originally belonged to the Indians, and were driving before them, as they advanced, immense herds of buffalo and fine ponies. The Great Spirit, who had so long deserted his red children, was now once more with them and against the whites, and the white man's gunpowder would no longer have power to drive a bullet through the skin of an Indian. The whites themselves would soon be overwhelmed and smothered under a deep landslide, held down by sod and timber, and the few who might escape would become small fishes in the rivers."[12]

James Mooney, an Irish American self-educated ethnographer, was assigned by John Wesley Powell, founder of the Bureau of American Ethnology, to investigate the Ghost Dance movement, traveling thirty-two thousand miles to visit twenty different tribes over twenty-two months. Mooney, an outspoken advocate of home rule for Ireland and later a cofounder of the Native American Church, saw the movement as an act of nonviolent self-preservation in the face of overwhelming odds. He traced its roots to the efforts of Tecumseh, the statesman-warrior, and Tenskwatawa, the prophet, to secure a homeland that had failed. "The wise men tell us that the world is growing happier—that we live longer than did our fathers, have more of comfort and less of toil, fewer wars and discords, and higher hopes and aspirations," begins his 634-page report. "So say the wise men; but deep in our own hearts we know they are wrong."[13]

Mooney gained sufficient trust among his subjects to participate in Ghost Dances with both the Arapahos and the Cheyennes before making a pilgrimage to meet Wovoka and learn about the origins of the movement firsthand. On his way to Nevada he made an extended visit to Pine Ridge, South Dakota, to report on the causes of the recent uprising among the Sioux and bear witness to the aftermath of the December 1890 massacre at Wounded Knee.

He gathered the opinions of eyewitnesses and officials ranging from Thomas J. Morgan, the commissioner of Indian affairs, to General Nelson Miles and Bishop W. H. Hare, Episcopal bishop to the Sioux. All agreed that the primary cause of the outbreak was not the Ghost Dance movement but starvation due to the disappearance of the buffalo combined with government promises to supply food that had gone unfulfilled.

Mooney concluded that "when the sun rose on Wounded Knee on the fatal morning of December 29, 1890, no trouble was anticipated or premeditated by either Indians or troops; that the Indians in good faith desired to surrender and be at peace, and that the officers in the same good faith had made preparations to receive their surrender and escort them quietly to the reservation; that in spite of the pacific intent of Big Foot and his band, the medicine man, Yellow Bird, at the critical moment urged the warriors to resistance and gave the signal for the attack; that the first shot was fired by an Indian, and that the Indians were responsible for the engagement; that the answering volley and attack by the troops was right and justifiable, but that the wholesale slaughter of women and children was unnecessary and inexcusable."[14]

John W. Comfort, of the First Artillery Division attached to the Seventh Cavalry, was one of the soldiers positioned with four Hotchkiss guns overlooking the Sioux encampment, "so as to be able to sweep it with canister if necessary . . . each Gunner training his P[ie]ce on a point in the Indian Camp where in his judgement it would do the most good."[15] The Hotchkiss was a new, forty-two-millimeter breech-loading rifled light mountain gun firing two-pound shells. The canister was loaded like a shotgun shell with smaller balls that spread out in flight and were indiscriminately lethal against any soft targets within range.

When fighting broke out within the encampment, the Hotchkiss gunners had to wait for the troopers to withdraw before "sending their Canister tearing into the mass of Warriors, Squaws, Children, Horses," at hundred-yard range. "It was a terrible thing to do," Comfort admitted, "but the Warriors were mixed with the Squaws, and were firing rapidly with their Winchesters, and Squaws were seen to kill Wounded Soldiers." Some 350 Sioux tried to escape, under fire from the Hotchkiss guns. "The Guns after exhausting their Canister used Percussion Shell continuing their fire until what was left of the flying Indians were hid from view in a ravine about 2500 Yd's distant," Comfort's account continues, "where three Troops of Cav[alry] rounded them up killing or capturing the entire party, nearly all the captured being wounded." A handful of survivors "g[a]ve the dismounted Troopers a lively fight until they were killed."[16]

Twenty-eight U.S. soldiers were killed, an unknown number of them by fire from their own side. Sioux deaths were tallied at 243 by Comfort and 185 by Joseph Horn Cloud, a survivor who listed their individual names.[17] They were buried without ceremony by army contractors in mass graves. "It was a thing to melt the heart of a man, if it was of stone," testified one of Mooney's informants, "to see those little children, with their bodies shot to pieces, thrown naked into the pit."[18]

Mooney arrived at the Walker River Agency, near Pyramid Lake, Nevada, on December 26, 1891, and found Wovoka to be forty-four miles away, in the upper reaches of the Mason valley, "a narrow strip of level sage prairie, some 30 miles in length, walled in by the giant sierras, their sides torn and gashed by volcanic convulsions and dark with gloomy forests of pine, their towering summits white with everlasting snows, and roofed over by a cloudless sky whose blue infinitude the mind instinctively seeks to penetrate to far-off worlds beyond."[19] C. C. Warner, the Indian agent at Pyramid Lake, assured Mooney that "there are neither ghost songs, dances, nor ceremonials among them about my agencies. Would not be allowed."[20] It took a full week for Mooney to gain the confidence of Wovoka's uncle, Charley Sheep, who agreed to set off in search of his nephew, whom they found, out hunting after a deep snowfall, on New Year's Day 1892. "Sure enough it was the messiah, hunting jack rabbits," Mooney reported. The deep snow was "a very unusual thing in this part of the country, and due in this instance, as Charley assured us, to the direct agency of Jack Wilson."[21]

In a dirt-floor tule-thatched hut, with his wife, infant, and four-year-old son in attendance, Wovoka welcomed his visitors into his home while tending a blazing sagebrush fire "upon which fresh stalks were thrown from time to time, sending up a shower of sparks into the open air."[22] His vision had appeared to him "when the sun died," three years earlier, corresponding to the records of a solar eclipse on January 1, 1889. Mooney attributed the vision to the effects of a high fever from an illness coupled with the excitement of the eclipse, and subsequent visions to a tendency to "cataleptic fits," while making careful notes of Wovoka's teachings, recording several of his songs, and gaining permission to take his photograph, seated in front of Charley Sheep with the fresh snow melting on the surrounding brush. He emphasized that Wovoka's message was one of peace, directing his followers to give up all violence and dishonesty, not only against the whites but also among themselves.

Wovoka claimed to have received these instructions, along with his powers to affect the weather, directly from God while he was unconscious during the eclipse. "I learned that Wovoka has five songs for making it rain," Mooney reported, "the first of which brings on a mist or cloud, the second a snowfall, the third a shower, and the fourth a hard rain or storm, while when he sings the fifth song the weather again becomes clear."[23]

When he compared notes with his Cheyenne and Arapaho informants who had also visited Wovoka, Mooney found their opinions differed. "Tall Bull, one of the Cheyenne delegates and then captain of the Indian police, said that before leaving they had asked Wovoka to give them some proof of his supernatural powers. Accordingly he had ranged them in front of him, seated on the ground, he sitting facing them, with his sombrero between and his eagle feathers in his hand. Then with a quick movement he had put his hand into the empty hat and drawn out from it 'something black.' Tall Bull would not admit that anything more had happened, and did not seem to be very profoundly impressed by the occurrence, saying that he thought there were medicine men of equal capacity among the Cheyenne."

"Black Coyote," one of the Arapaho delegates, "told how they had seated themselves on the ground in front of Wovoka, as described by Tall Bull, and went on to tell how the messiah had waved his feathers over his hat, and then, when he withdrew his hand, Black Coyote looked into the hat and there 'saw the whole world.'"

"When the messiah performed his hypnotic passes with the eagle feather, as I have so often witnessed in the Ghost dance," Mooney concluded, without passing judgment, "Black Coyote saw the whole spirit world where Tall Bull saw only an empty hat."[24]

In the twentieth century, digital computers advanced across North America, in the aftermath of World War II, as inexorably as the railroads had advanced across the plains in the aftermath of the Civil War in the nineteenth. A series of lone voices raised the alarm, led by Norbert Wiener, the co-founder, with Julian Bigelow, of modern cybernetics, beginning in 1943 with their prophetic "Behavior, Purpose, and Teleology,"[25] and ending in 1964, the year of Wiener's death, with a prediction that "the world of the future will be an ever more demanding struggle against the limitations of our intelligence, not a comfortable hammock in which we can lie down to be waited upon by our robot slaves."[26] Those who seek to become minders over robots may end up minded by robots, instead.

These warnings were dismissed, first of all because the development of digital computers was accompanied by the development of the hydrogen bomb, whose near-term dangers eclipsed the long-term dangers of relinquishing human agency to machine control. Second, how could the autonomy of humans be at risk, as long as we are supplying the instructions to the machines?

That digital computers are dependent on algorithms, or logical step-by-step procedures, is no guarantee that machine intelligence will remain under logical control. Expecting to find an algorithm underlying machine intelligence may be as futile as expecting to find a language underlying communication among whales. The domain of algorithms may be the wrong place to look for signs of true autonomy and intelligence among machines.

Nature evolved analog computers, known as nervous systems, that embody information absorbed from the world. They learn. They learn to control their own behavior, and they learn to control their environment, including the behavior of other organisms of their own and of other kinds. "What makes you so sure," asked Stan Ulam, more than three generations of machine intelligence ago, "that mathematical logic corresponds to the way we think?"[27]

———

On March 25, 1943, two days after Freeman Dyson bicycled from Cambridge to Ely, encountering a fleet of RAF bombers on the way, he received "a large envelope stamped Princeton, Feb. 11, 1943, and inside it, lo and behold, 'The Consistency of the Continuum Hypothesis,' by K[urt] Gödel. This is the first time I have ever been aware," he reported to his parents, "except from an abstract point of view, that a place called America really exists."[28]

The continuum hypothesis, originating in Leibniz's work on infinitesimals, suggests that analog computing and digital computing, each with unlimited powers, will continue to exhibit *different* powers, no matter how far they advance. As first articulated by Georg Cantor, the pioneer of transfinite numbers, in 1878, the continuum hypothesis divides all infinities, of which there are an infinite number, into two distinct types. On one side is the infinite set of whole numbers constructed by counting the integers 1, 2, 3 . . . without end. On the other side is the infinite set of real numbers represented by the points on a line. Not only are the points on any line, even of finite length, of infinite number, but between any two points, no matter how close, there are also an infinite number of points.

Imagine a wide, empty beach, piled deep with perfect grains of sand, stretching into the distance without end. At the edge of the water is drawn a single continuous line. Because the beach goes on forever, there are an infinite number of grains of sand, but it is a countable infinity, because you could count the grains of sand one by one. There are also an infinite number of points on the line, but because the line is a continuum, it is an uncountable infinity. If you sat down with a friend on the beach and started counting the points on the line while your friend started counting the grains of sand, you would either have to skip an uncountably infinite number of points to keep up, or your friend would leave you behind. The two infinities will never match up.

According to the continuum hypothesis, all countable infinities, like the number of discrete grains of sand, can be placed in a one-to-one correspondence with the integers. All uncountable infinities, like the number of points on a continuous line, have the full power of the continuum. There are no infinities in between. Infinities are like free T-shirts at the end of a conference: available in only extra large and extra small. If the continuum hypothesis is true, infinities come in only two sizes, or powers, and no medium-sized infinities will ever be found. At the heart of the continuum hypothesis is the conjecture that there is a fundamental difference, leaving no middle ground, between infinities that are continuous, or uncountable, and infinities that are countable, or discrete.

Cantor credited the origins of the hypothesis to Leibniz, while Gödel, who devoted years to the problem, in his words, whether "every infinite subset of the continuum has either the power of the set of integers or of the whole continuum," searched for clues among Leibniz's unpublished writings for much of his later life.[25] Some of his IAS colleagues felt he was wasting his time, and diminishing the mathematical standards of the institute, over his obsession with the Leibniz manuscripts, but von Neumann came to his defense, arguing that Gödel "ought to be the sole judge of what he does."[30] It may be that Gödel, whose insights, as much as Alan Turing's or John von Neumann's, underlie the powers of digital computing, was, in his obsession over the continuum hypothesis and Leibniz's unfinished agenda, on the track of powers that lie beyond.

In 1900, David Hilbert placed the continuum hypothesis, "a very plausible theorem, which nevertheless, in spite of the most strenuous efforts, no one has succeeded in proving," at the top of his list of twenty-three problems that were unsolved.[31] It remains unsolved, although Gödel, in 1940, by invoking a "constructible" universe, proved its independence from the axioms of set theory, meaning that it could be true, even if unprovable, and Paul Cohen, in 1963, went further in proving that without additional axioms it cannot be proved. Truth and provability are two different things.[32]

There is a corollary to the continuum hypothesis concerning computation among living and nonliving things. Computers, like Cantor's infinities, can be divided into two kinds. Digital computers are finite but unbounded discrete-state machines whose possible states can be mapped in one-to-one correspondence to the integers. Analog computers, lacking discrete states that can be mapped directly to the integers, belong instead to some subset of the continuum, with every such subset, according to Cantor, having the power of the continuum as a whole.

Digital computers deal with integers, binary sequences, deterministic logic, and time that is idealized into discrete increments. Analog computers deal with real numbers, nondeterministic logic, and continuous functions, including time as it exists as a continuum in the real world.

Imagine you need to find the middle of a road. You can measure its width using any available increment and then digitally compute the middle to the nearest increment. Or you can use a length of string as an analog computer, mapping the width of the road to the length of the string and finding the middle, without being limited to increments, by doubling the string back upon itself.

In analog computing, complexity resides in architecture, not code. Information is processed as continuous functions of values such as voltage and relative pulse frequency rather than by logical operations on discrete sequences of bits. To an analog computer, one bit here or there makes no great difference. To a digital computer, one bit can make all the difference in the world.

Digital computing, intolerant of error or ambiguity, depends upon precise definitions and error correction at every step. Analog computing not only tolerates errors and ambiguities but learns to thrive on them. Digital computers, in a technical sense, are analog computers so hardened against noise that they have lost their immunity to it. Analog computers *embrace* noise: a real-world neural network, such as the visual or auditory system in a developing brain, requiring a certain level of background noise in order to work.

Nature uses a quaternary alphabet of nucleotides to store, replicate, and transmit an unbounded library of instructions for the reproduction of otherwise non-digital living things, coded in a way that is optimized for modification, recombination, and error correction along the way. Incorporating both the countable and the uncountable, Nature uses digital computing for generation-to-generation information storage, combinatorics, and error correction but relies on analog computing for real-time intelligence and control.

The constructible infinities are currently, but not necessarily, favored by machines. The continuum hypothesis suggests there are no intermediate powers in between. No matter how large, how fast, and how powerful, digital computers remain limited to the powers of the denumerable, and, no matter how small or insignificant, organisms, and machines that operate like organisms, share their powers with the continuum as a whole.

Analog computing is alive and well despite vacuum tubes being commercially extinct. It is advancing on two fronts: from the bottom up, driven by drone warfare, autonomous vehicles, and mobile devices pushing the development of neuromorphic, or neuron-like, microprocessors; and from the top down as networks formed by interconnected populations of digital computers turn to analog computation in their infiltration and control of the world. These systems are supervening upon the digital, forming a new computational layer in the same way that digital computers were coaxed into existence as a new layer of abstraction running on analog hardware in the aftermath of World War II.

Individually deterministic finite-state processors, running finite codes, are forming large-scale, nondeterministic, non-finite-state metazoan systems that treat streams of bits collectively, the way the flow of electrons is treated in a vacuum tube, rather than individually, as bits are treated by the discrete-state devices generating the flow. Bits are the new electrons. Governing everything from the flow of goods to the flow of traffic to the flow of ideas, information is treated statistically, the way pulse-frequency-coded information is processed in a neuron or a brain. Analog is back, and its nature is to assume control.

Imagine it is 1958 and you want to defend North America against airborne attack. To distinguish hostile aircraft, you need a network of computers and early warning radar sites to identify intruders, checked against a map of all commercial air traffic, updated from one minute to the next. The United States built such a system and named it SAGE (Semi-Automatic Ground Environment). SAGE in turn spawned SABRE, the first integrated reservation system for booking airline travel online. SABRE and its progeny soon became not just a map as to what seats were available on which flights but also a system that began to control, with decentralized intelligence, where airliners would fly and when.

The airlines called this "yield management" and developed sophisticated algorithms for changing the price of tickets according to available seats in real time. But the essential intelligence, and the control, aren't in the coded algorithms; they are in the network that is mapped to the flights. Ticket price and flight frequency are two of the continuous functions that are computed by the network, and those signals end up controlling not just the ticket price and the frequency of flights but the geography of the map.

Isn't there a control room somewhere with someone at the controls? Maybe not. Say, for example, you build a system to map highway traffic in real time, by giving cars access to the map in exchange for reporting their own location and speed. The result is a decentralized control system. Nowhere is there any central controlling model besides the real-time reports, a real-time map, and a simple algorithm that chooses the shortest path in time between two points. The complexity is not in the algorithm; it's in the traffic itself.

This is how social networks gain power and why they scale so well. They run on digital computers, but it is the statistical mapping of relations between users, not the underlying logical code, that counts. A model of the social graph *becomes* the social graph, and proliferates across the world.

What if you wanted to capture what everything known to the human species *means*? Thanks to Moore's law it takes little time and less and less money to capture all the information that exists. But how do you capture meaning?

Even in the age of all things digital, this cannot be defined in logical terms, because meaning, among humans, isn't logical. Leibniz's logical utopia fails to close. The best you can do, once you have collected all possible answers, is to invite well-defined questions and compile a pulse-frequency-weighted map of how everything connects. This system, in conjunction with illogical humans, will not only be observing and mapping the meaning of things; it will start constructing meaning as well, the way a dictionary doesn't just catalog a language, but defines the language, over time. The meaning of something is established, among humans, by the degree to which that something connects to other familiar things. A search engine, mapping those connections, isn't just a collective model of how we think; increasingly, it *is* how we think. In time it will *control* meaning, in the same way as the traffic map controls the flow of traffic, even though no one is in control.

There are three laws of artificial intelligence. The first, known as Ashby's law of requisite variety after the cybernetician W. Ross Ashby, author of *Design for a Brain*, states that any effective control system must be as complex as the system it controls.

The second law, articulated by John von Neumann, states that the defining characteristic of a complex system is that it constitutes its own simplest behavioral description. The simplest complete model of an organism is the organism itself. Trying to reduce the system's behavior to a formal description, such as an algorithm, makes things more complicated, not less.

The third law states that any system simple enough to be understandable will not be complicated enough to behave intelligently, while any system complicated enough to behave intelligently will be too complicated to understand.

These laws seem to imply that artificial intelligence capable of thinking for itself will never be reached through formally programmable control. They offer comfort to those who believe that until we understand human intelligence, we need not worry about superhuman intelligence among machines. But there is no law against building something *without* understanding it.

During the middle epochs of technology, the powers of the continuum were left to nature, while the powers of the countable infinities were exercised by machines. The absence of medium-sized infinities left a vacuum for self-reproducing technology and self-replicating codes to fill. The number of bits in the digital universe is countable, but growing so quickly that if you stop to sample any subset, you will always find an increasing number of bits. It is as if you were back on that beach counting the points on the line, but the number of grains of sand is doubling so fast you are no longer completely outpaced by your friend who is counting the grains of sand. If ever there was a medium-sized infinity, it would look like this.

The digital universe is currently expanding by some thirty trillion transistors per second, populated by unbounded, if countable, strings of code. This proliferation is driven by three underlying principles: Turing's demonstration of universality among digital computers, realizing Leibniz's vision of a universal language common to all machines; von Neumann's theory of self-reproducing automata, proving that any such universal machine can reproduce itself; and Shannon's sampling theorem, showing how any continuous function can be captured with arbitrary fidelity by a discrete-state machine, paving the way for digital computers to dominate the world.

In the fourth epoch of technology, the powers of the continuum will be claimed by machines. The next revolution, as fundamental as when analog components were assembled into digital computers, will be the rise of analog systems over which the dominion of digital programming comes to an end. Nature's answer to those who seek to control nature through programmable machines is to allow us to build systems whose nature is beyond programmable control.

———

Nothing is to be gained by resisting the advance of the discrete-state machines, for the ghosts of the continuum will soon return, when the grass is eight inches high in the spring. A world shared with true machine intelligence, proliferating in the wild as opposed to the captive, domesticated versions being marketed today, will bring the new millennium that Wovoka and others predicted, infused by an intelligence that belongs not to us but to Nature as a whole. The human species, and the human mind, grew up in a world shared with creatures large and small, animated by spirits that some were privileged to communicate with but no one claimed to understand, let alone control. "The least particle must be considered as a world full of an infinity of different creatures," Leibniz admitted in 1693.[33]

Some see only an empty hat. Others see the full power of the continuum, with the powers of the merely countable infinities becoming small fishes in the rivers while megafauna once again roam the earth.

Not even ashes will remain.

END

NOTES

The views of the Northwest Coast of America appearing across these pages were engraved on copper after drawings made in 1778 by John. Webber and William Bligh during Cook's third voyage, and printed in volume 2 of James Cook and James King, *A Voyage to the Pacific Ocean, Undertaken, by the Command of His Majesty, for Making Discoveries in the Northern Hemisphere, to Determine the Position and Extent of the West Side of North America; Its Distance from Asia; and the Practicability of a Northern Passage to Europe, Performed Under the Direction of Captains Cook, Clerke, and Gore, in His Majesty's Ships the* Resolution *and* Discovery, *in the Years 1776, 1777, 1778, 1779, and 1780*, published by order of the lord commissioners of the Admiralty in 1784.

All illustrations in the text and the illustration inserts are from the author's collection, and all photographs are by the author, unless otherwise noted.

Key to archival repositories:

AFSWC: U.S. Air Force Special Weapons Center, History Office, Kirtland Air Force Base, Albuquerque, N.M.

CBI: Charles Babbage Institute, University of Minnesota, Minneapolis.

FJD: Freeman Dyson Papers, American Philosophical Society, Philadelphia.

GA: General Atomic (now General Atomics), La Jolla, Calif.

GBD: Author's personal papers. Interviews with the author unless otherwise noted.

IAS: Shelby White and Leon Levy Archives Center, Institute for Advanced Study, Princeton, N.J.

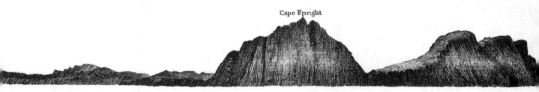

View of Gores Island when Cape Upright bore NN. W. 3 miles dist.

O. THE LEIBNIZ ARCHIPELAGO

1. E. T. Bell, *Men of Mathematics* (New York: Simon & Schuster, 1937), 122.
2. Gottfried Wilhelm Leibniz (ca. 1679), *Philosophical Papers and Letters*, trans. and ed. Leroy E. Loemker (Chicago: University of Chicago Press, 1956), 1:344.
3. Ibid.
4. Vladimir Guerrier, *Leibniz in seinen Beziehungen zu Russland und Peter dem Grossen* (St. Petersburg: Kaiserlichen Akademie der Wissenschaften, 1873), 1:141. Author's translation.
5. John von Neumann, "The General and Logical Theory of Automata," September 20, 1948, in *Cerebral Mechanisms in Behavior: The Hixon Symposium*, ed. Lloyd A. Jeffress (New York: Hafner, 1951), 21.
6. Carver Mead, *Analog VLSI and Neural Systems* (Reading, Mass.: Addison-Wesley, 1989), xi.

I. 1741

1. In Russia, dates were recorded in the Julian calendar, eleven days behind the Gregorian calendar, in 1741. Additionally, vessel logs were kept according to the astronomical date, advancing at noon instead of midnight, so entries before noon agree with the calendar date, but afternoon entries are one day ahead.
2. Sven Waxell, *The American Expedition*, trans. M. A. Michael (London: William Hodge, 1952), 128.
3. Georg Wilhelm Steller, *Journal of a Voyage with Bering, 1741–1742*, ed. O. W. Frost, trans. Margritt A. Engel and O. W. Frost (Stanford, Calif.: Stanford University Press, 1988), 93.
4. Thomas Consett, *An Account of the Rise of Naval Power in Russia; or, The Story of the Little Boat, Which Gave Rise to the Russian Fleet; Being the Preface to the Sea Regulations in Russia, and Said to Have Been Written by the Tsar Peter Alexievich Himself*, in *The Present State and Regulations of the Church of Russia* (London: S. Holt, 1729), 2:209.
5. John Evelyn, January 30, 1698, and an undated note from his servant, in *The Diary of John Evelyn*, ed. William Bray (Washington, D.C.: Walter Dunne, 1901), 2:342.
6. Thomas Hale to Bernard Hale, August 20, 1702 (British Museum Add. Mss. 33573, fol. 178), in James Cracraft, *The Church Reform of Peter the Great* (London: Macmillan, 1971), 10.
7. Leibniz to Bernoulli, July 26, 1716, in Vladimir Guerrier, *Leibniz in seinen Beziehungen zu Russland und Peter dem Grossen* (St. Petersburg: Kaiserlichen Akademie der Wissenschaften, 1873), 2:361. Translation by Scott Sessions.
8. Leibniz to Sophie von Hannover, November 1712, in Guerrier, *Leibniz in seinen Beziehungen zu Russland und Peter dem Grossen*, 2:272. Author's translation.
9. Instructions from Empress Catherine Alekseevna to Captain Vitus Bering for the First Kamchatka Expedition, February 25, 1725 (written by Peter the Great before his death on January 28, 1725), in *Russian Penetration of the North Pacific Ocean, 1700–1797: A Documentary Record*, ed. Basil Dmytryshyn, E. A. P. Crownhart-Vaughan, and Thomas Vaughan (Portland: Oregon Historical Society, 1988), 69.
10. Peter Lauridsen, *Vitus Bering: The Discoverer of Bering Strait*, trans. Julius E. Olson (Chicago: S. C. Griggs, 1889), 69.

11. Waxell, *American Expedition*, 51.
12. Ibid., 67–68.
13. *A Letter from a Russian Sea-Officer, to a Person of Distinction at the Court of St. Petersburgh* (London: A. Linde, 1754), 21–22. Attributed to G. F. Müller, and translated from the French of 1753.
14. Georg Wilhelm Steller, in *Bering's Voyages: An Account of the Efforts of the Russians to Determine the Relation of Asia and America*, ed. F. A. Golder, vol. 2, *Steller's Journal of the Sea Voyage from Kamchatka to America and Return on the Second Expedition, 1741–1742* (New York: American Geographical Society, 1925), 48.
15. Steller, *Journal of a Voyage with Bering*, 72.
16. Steller, *Journal of the Sea Voyage from Kamchatka to America*, 60.
17. Ibid., 54.
18. Waxell, *American Expedition*, 98.
19. Steller, *Journal of the Sea Voyage from Kamchatka to America*, 68.
20. Ibid., 90.
21. Ibid., 90–92.
22. Ibid., 94.
23. Sofron Khitrov, in *Bering's Voyages: An Account of the Efforts of the Russians to Determine the Relation of Asia and America*, ed. F. A. Golder, vol. 1, *The Log Books and Official Reports of the First and Second Expeditions, 1725–1730 and 1733–1742* (New York: American Geographical Society, 1922), 148.
24. Steller, *Journal of the Sea Voyage from Kamchatka to America*, 95.
25. Knut R. Fladmark, "The Feasibility of the Northwest Coast as a Migration Route for Early Man," in *Early Man in America, from a Circum-Pacific Perspective*, ed. Bryan Alan Lyle (Edmonton: Archaeological Researches International, 1978), 119–28; Jon M. Erlandson et al., "The Kelp Highway Hypothesis: Marine Ecology, the Coastal Migration Theory, and the Peopling of the Americas," *Journal of Island and Coastal Archaeology* 2 (2007): 161–74.
26. Laughlin, interview, Umnak Island, 1991, GBD.
27. Waxell, *American Expedition*, 117.
28. Golder, *Bering's Voyages*, 1:167.
29. Steller, *Journal of a Voyage with Bering*, 115.
30. Waxell, *American Expedition*, 199, 121.
31. Golder, *Bering's Voyages*, 1:196, 200.
32. Steller, *Journal of the Sea Voyage from Kamchatka to America*, 129–30.
33. Waxell, *American Expedition*, 124.
34. Steller, *Journal of a Voyage with Bering*, 125.
35. Waxell, *American Expedition*, 125.
36. Steller, *Journal of a Voyage with Bering*, 126–27.
37. Steller, *Journal of the Sea Voyage from Kamchatka to America*, 136.
38. Steller, *Journal of a Voyage with Bering*, 127.
39. Waxell, *American Expedition*, 135.
40. Ibid., 127, 135.
41. Steller, *Journal of a Voyage with Bering*, 136.
42. Steller to Gmelin, November 4, 1742, in Golder, *Bering's Voyages*, 2:243.
43. Waxell, *American Expedition*, 135.

44. Ibid., 142.
45. Steller, *Journal of the Sea Voyage from Kamchatka to America*, 148.
46. Ibid., 149.
47. George Wilhelm Steller, *The Beasts of the Sea*, trans. Walter Miller and Jennie Emerson Miller, in *The Fur Seals and Fur-Seal Islands of the North Pacific Ocean*, ed. David Starr Jordan, pt. 3 (Washington, D.C.: Government Printing Office, 1899), 216.
48. Ibid., 215.
49. Steller, *Journal of a Voyage with Bering*, 164.
50. Steller, *Beasts of the Sea*, 197.
51. Ibid., 201.
52. Ibid., 191.
53. Ibid., 200.
54. Ibid., 164.
55. Gerhard Friedrich Müller, *Bering's Voyages: The Reports from Russia*, trans. Carol Urness (Fairbanks: University of Alaska Press, 1986), 120.
56. Gerhard F. Müller, *Voyages from Asia to America* (London: T. Jeffreys, 1761), 65.
57. Georg Wilhelm Steller, August 18, 1746, in Leonhard Stejneger, *Georg Wilhelm Steller: The Pioneer of Alaskan Natural History* (Cambridge, Mass.: Harvard University Press, 1936), 476.
58. Steller, *Beasts of the Sea*, 181.
59. Ibid., 181–82.
60. Don Douglass, "Possible Clues to Chirikov's Lost Crewmen" (presentation for 2010 International Conference on Russian America, Sitka, Alaska, August 19–21, 2010).
61. Alexei Chirikov, *Report on the Voyage of the* St. Paul, December 7, 1741, in Golder, *Bering's Voyages*, 1:315.
62. Aleksei Chirikov, *Report to the Admiralty College*, December 7, 1741, in *Anóoshi Lingít Aaní Ká, Russians in Tlingit America*, ed. Nora Marks Dauenhauer, Richard Dauenhauer, and Lydia T. Black (Seattle: University of Washington Press, 2008), 8.
63. Alexei Chirikov, *Journal of the* St. Paul, July 24, 1741, in Golder, *Bering's Voyages*, 1:295.
64. Chirikov, *Report on the Voyage of the* St. Paul, December 7, 1741, 316.
65. Ibid., 316–17.
66. J.F.G. de La Pérouse, *A Voyage Round the World, Performed in the Years 1785, 1786, 1787, and 1788, by the* Boussole *and* Astrolabe (London: G. G. & J. Robinson, 1799), 390, 400.
67. Chirikov, *Report on the Voyage of the* St. Paul, December 7, 1741, 315.
68. Lydia T. Black, notes to a new translation of Chirikov, *Report to the Admiralty College, December 7, 1741*, 9.
69. Chirikov, *Report on the Voyage of the* St. Paul, December 7, 1741, 317.
70. Ibid.
71. Mark Jacobs Jr., "Early Encounters Between the Tlingit and the Russians," in *Russia in North America: Proceedings of the 2nd International Conference on Russian America, Sitka, Alaska, August 19–22, 1987*, ed. Richard A. Pierce (Kingston, Ont.: Limestone Press, 1990), 2.
72. Chirikov, *Journal of the* St. Paul, September 9, 1741, in Golder, *Bering's Voyages*, 1:304–305.

2. LAST OF THE APACHES

1. Nelson A. Miles, "The Future of the Indian Question," *North American Review* 152, no. 410 (January 1891): 2.

2. Geronimo, *Geronimo's Story of His Life*, ed. S. M. Barrett (New York: Duffield, 1906), 178. Field translation by Asa "Ace" Daklugie.

3. Britton Davis, *The Truth About Geronimo* (New Haven, Conn.: Yale University Press, 1929), 17.

4. Asa "Ace" Daklugie, in Eve Ball, *Indeh: An Apache Odyssey* (Provo, Utah: Brigham Young University Press, 1980), 55.

5. Thomas Cruse, *Apache Days and After* (Caldwell, Idaho: Caxton Printers, 1941), 164.

6. Nelson A. Miles, *Personal Recollections and Observations of General Nelson A. Miles, Embracing a Brief View of the Civil War; or, From New England to the Golden Gate, and the Story of His Indian Campaigns, with Comments on the Exploration, Development, and Progress of Our Great Western Empire* (Chicago: Werner, 1896), 431.

7. Philip Reade, "About Heliographs," *United Service, a Monthly Review of Military and Naval Affairs*, January 1880, 108.

8. G. W. Baird, "General Miles's Indian Campaigns," *Century Magazine*, July 1891, 368.

9. General Nelson A. Miles, September 17, 1886, in *Letter from the Secretary of War, February 28, 1887, Executive Document No. 117 of the Senate of the United States for the Second Session of the Forty-Ninth Congress* (Washington, D.C.: Government Printing Office, 1887), 43.

10. Álvar Núñez Cabeza de Vaca (1536), *The Journey of Álvar Núñez Cabeza de Vaca and His Companions from Florida to the Pacific, 1528–1536*, ed. Adolph F. Bandelier, trans. Fanny Bandelier (New York: Barnes, 1905), 171.

11. Ibid., 183.

12. Pedro de Castañeda de Nájera, in *Documents of the Coronado Expedition, 1539–1542: "They Were Not Familiar with His Majesty, nor Did They Wish to Be His Subjects,"* ed. Richard Flint and Shirley Cushing Flint (Dallas: Southern Methodist University Press, 2005), 387–88.

13. Pedro de Castañeda de Nájera, in *The Coronado Expedition, 1540–1549*, ed. George Parker Winship, in *The Fourteenth Annual Report of the Bureau of Ethnology* (Washington, D.C.: Government Printing Office, 1896), 516.

14. Francisco Vázquez de Coronado, in Winship, *Coronado Expedition*, 556.

15. Ibid., 557, 559, 565.

16. Ibid., 555.

17. Coronado to the Viceroy, August 3, 1540, in Flint and Flint, *Documents of the Coronado Expedition*, 257.

18. Coronado, in Winship, *Coronado Expedition*, 563.

19. Antonio de Mendoza to the King of Spain, April 17, 1540, in Flint and Flint, *Documents of the Coronado Expedition*, 238.

20. Baylor to Thomas Helm, March 20, 1862, in *The War of the Rebellion: A Compilation of the Official Records of the Union and Confederate Armies* (Washington, D.C.: Government Printing Office, 1896), 50:942.

21. Sibley to Adjutant General S. Cooper, May 4, 1862, in *The War of the Rebellion: A Compilation of the Official Records of the Union and Confederate Armies* (Washington, D.C.: Government Printing Office, 1883), 9:512.

22. Carleton to Carson, October 12, 1862, in U.S. Senate, 39th Cong., 2nd Sess., Report No. 156, *Condition of the Indian Tribes* (Washington, D.C.: Government Printing Office, 1867), app., p. 100.

23. Daniel Ellis Conner, *Joseph Reddeford Walker and the Arizona Adventure*, ed. Donald J. Berthrong and Odessa Davenport (Norman: University of Oklahoma Press, 1956), 36.

24. Ibid., 38–39.

25. Orson S. Fowler, *Human Science, or, Phrenology: Its Principles, Proofs, Faculties, Organs, Temperaments, Combinations, Conditions, Teachings, Philosophies, etc., etc.: As Applied to Health, Its Values, Laws, Functions, Organs, Means, Preservation, Restoration, etc.: Mental Philosophy, Human and Self Improvement, Civilization, Home, Country, Commerce, Rights, Duties, Ethics, etc.: God, His Existence, Attributes, Laws, Worship, Natural Theology, etc.: Immortality, Its Evidences, Conditions, Relations to Time, Rewards, Punishments, Sin, Faith, Prayer, etc.: Intellect, Memory, Juvenile and Self Education, Literature, Mental Discipline, the Senses, Sciences, Arts, Avocations, a Perfect Life, etc., etc., etc.* (Philadelphia: National, 1873), 1195, 1197.

26. Davis, *Truth About Geronimo*, 59.

27. Daklugie, in Ball, *Indeh*, 89, 15.

28. Ibid., 37.

29. James Kaywaykla, in Eve Ball, *In the Days of Victorio* (Tucson: University of Arizona Press, 1970), xiii–xiv.

30. John Gregory Bourke, *On the Border with Crook* (New York: Scribner's, 1891), 127, 129.

31. Davis, *Truth About Geronimo*, 114.

32. Crook to the Secretary of War, September 27, 1883, in *Annual Report of the Secretary of War for the Year 1883* (Washington, D.C.: Government Printing Office, 1883), 1:67.

33. George Crook, *Resumé of Operations Against Apache Indians, 1882 to 1886* (Omaha: privately printed, 1886), 21.

34. Tom Horn, *Life of Tom Horn, Government Scout and Interpreter, Written by Himself* (Denver: Louthan, 1904), 18.

35. Alchisay, testimony at Fort Apache, September 22, 1882, in Bourke, *On the Border with Crook*, 436.

36. "Transcript of Conference Between General Crook and Between 400 and 500 Men of the Apache Tribe at San Carlos Agency, Ariz., October 15, 1882," in *Annual Report of the Secretary of War for the Year 1883*, 1:180.

37. *Annual Report of the Secretary of War for the Year 1883*, 1:182.

38. Davis, *Truth About Geronimo*, 103, 111.

39. Geronimo, *Geronimo's Story of His Life*, 17.

40. Ibid., 38–39.

41. Ibid., 141.

42. *Notes of an Interview Between Maj. General George Crook, U.S. Army, and Chatto, Ka-e-te-na, Noche, and Other Chiricahua Apaches, January 2, 1890*, in U.S. Senate, 51st Cong., 1st Sess., Ex. Doc. No. 35 (Washington, D.C.: Government Printing Office, 1890), 5.

43. Jason Betzinez, *I Fought with Geronimo* (Harrisburg, Pa.: Stackpole Books, 1959), 137.

44. Dan L. Thrapp, *Encyclopedia of Frontier Biography* (Glendale, Calif.: Arthur Clark, 1988), 2:914.

45. Crook, *Resumé of Operations Against Apache Indians*, 9.

46. Bourke, *On the Border with Crook*, 477.

47. Charles Lummis, *Los Angeles Times*, April 11, 1886, in *Dateline Fort Bowie: Charles Fletcher Lummis Reports on an Apache War*, ed. Dan L. Thrapp (Norman: University of Oklahoma Press, 1979), 57–58.

48. Sheridan to Crook, February 1, 1886, in Crook, *Resumé of Operations Against Apache Indians*, 8.

49. Crook to Sheridan, March 26, 1886, in Crook, *Resumé of Operations Against Apache Indians*, 10.

50. Geronimo, March 27, 1886, in *Conference Held March 25 and 27, 1886, at Cañon de los Embudos (Cañon of the Funnels), 20 Miles S.SE. of San Bernardino Springs, Mexico, Between General Crook and the Hostile Chiricahua Chiefs*, in U.S. Senate. 51st Cong., 1st Sess., Ex. Doc. No. 88 (Washington, D.C.: Government Printing Office, 1890), 11.

51. Sheridan to Crook, March 30, 1886, in Crook, *Resumé of Operations Against Apache Indians*, 12.

52. Bourke, *On the Border with Crook*, 480.

53. *Notes of an Interview Between Maj. General George Crook, U.S. Army, and Chatto, Ka-e-te-na, Noche, and Other Chiricahua Apaches, January 2, 1890*, 5.

54. Daklugie, in Ball, *Indeh*, 62.

55. Cleveland to General R. C. Drum, August 23, 1886, in *Letter from the Secretary of War, February 28, 1887*, 4.

56. Miles, *Personal Recollections*, 136.

57. Reade, "About Heliographs," 94.

58. William W. Neifert, "Trailing Geronimo by Heliograph," in *Winners of the West* 12, no. 11 (October 1935), in *The Struggle for Apacheria*, vol. 1 of *Eyewitnesses to the Indian Wars, 1865–1890*, ed. Peter Cozzens (Mechanicsburg, Pa.: Stackpole Books, 2001), 561.

59. Miles, *Personal Recollections*, 480.

60. Ibid., 487.

61. Ibid., 487–88.

62. Leonard Wood, August 6, 1886, in *Chasing Geronimo: The Journal of Leonard Wood, May–September 1886*, ed. Jack C. Lane (Albuquerque: University of New Mexico Press, 1970), 92.

63. Leonard Wood, August 4, 1886, in *Chasing Geronimo*, 89.

64. Henry Daly, "The Geronimo Campaign," *Journal of the United States Cavalry Association* 19, no. 70 (October 1908), in Cozzens, *Struggle for Apacheria*, 483.

65. Charles Gatewood, in *Lt. Charles Gatewood and His Apache War Memoir*, ed. Louis Kraft (Lincoln: University of Nebraska Press, 2005), 136.

66. James R. Caffey, "A Theatrical Campaign," *Omaha Bee*, September 29, 1886, in Cozzens, *Struggle for Apacheria*, 564.

67. Miles to Adjutant-General R. C. Drum, September 17, 1886, in *Letter from the Secretary of War, February 28, 1887*, 6.

68. Sheridan to the Secretary of War, July 30, 1886, in *Letter from the Secretary of War, February 28, 1887*, 52.

69. Daklugie, in Ball, *Indeh*, 113–14.

70. Colonel William Strover, "The Last of Geronimo and His Band," *Washington National Tribune*, July 24, 1924.

71. Crook to the Secretary of War, January 6, 1890, in U.S. Senate, 51st Cong., 1st Sess., Ex. Doc. No. 35 (Washington, D.C.: Government Printing Office, 1890), 5.

72. Kaywaykla, in Ball, *In the Days of Victorio*, 197.
73. Eugene Chihuahua, in Ball, *Indeh*, 113.
74. General Nelson Miles, November 4, 1886, in *Letter from the Secretary of War, February 28, 1887*, 77.
75. Daklugie, in Ball, *Indeh*, 118.

3. AGE OF REPTILES

1. Thomas A. Edison, "Electrical Indicator," U.S. Patent No. 307,031, application filed November 15, 1883, issued October 21, 1884.
2. William Henry Preece, "On a Peculiar Behaviour of Glow-Lamps When Raised to High Incandescence," *Proceedings of the Royal Society of London*, March 26, 1885, 219.
3. John Ambrose Fleming, "The Thermionic Valve in Wireless Telegraphy and Telephony," Royal Institution of Great Britain, Weekly Evening Meeting, May 21, 1920, 2.
4. W. H. Eccles, "John Ambrose Fleming, 1849–1945," *Obituary Notices of Fellows of the Royal Society* 5, no. 14 (November 1945): 231.
5. John Ambrose Fleming, *The Origin of Mankind: Viewed from the Standpoint of Revelation and Research* (London: Marshall, Morgan & Scott, 1935), v–vi.
6. Ibid., 7.
7. John Ambrose Fleming, "Problems in the Physics of an Electric Lamp," Royal Institution of Great Britain, Weekly Evening Meeting, February 14, 1890, 2, 4, 9.
8. J. A. Fleming, "A Further Examination of the Edison Effect in Glow Lamps," *Philosophical Magazine*, July 1896, 102.
9. Degna Marconi, *My Father, Marconi* (New York: McGraw-Hill, 1962), 13.
10. Flood-Page to Fleming, December 1, 1900, in Sungook Hong, *Wireless: From Marconi's Black-Box to the Audion* (Cambridge, Mass.: MIT Press, 2001), 69.
11. Marconi to Fleming, December 10, 1900, in Hong, *Wireless*, 69.
12. *Daily Telegraph*, December 18, 1901, in Marconi, *My Father, Marconi*, 116.
13. John Ambrose Fleming, "Improvements in Instruments for Detecting and Measuring Alternating Currents," British Patent No. 24,850, application filed November 16, 1904, accepted September 21, 1905.
14. Fleming to Marconi, November 30, 1904, Marconi Company Archive.
15. Fleming, "Thermionic Valve in Wireless Telegraphy and Telephony," 7.
16. John Ambrose Fleming, *The Thermionic Valve and Its Developments in Radiotelegraphy and Telephony* (London: Wireless Press, 1919), 48.
17. Georgette Carneal, *A Conqueror of Space: An Authorized Biography of the Life and Work of Lee De Forest* (New York: Horace Liveright, 1930), 95.
18. Lee de Forest, "Oscillation-Responsive Device," U.S. Patent No. 979,275, application filed February 2, 1905, issued December 20, 1910.
19. Lee de Forest, "The Audion—Its Action and Some Recent Applications," *Journal of the Franklin Institute* 190, no. 1 (July 1920): 2.
20. Lee de Forest, "Space Telegraphy," U.S. Patent No. 879,532, application filed January 29, 1907, issued February 18, 1908.
21. Ibid.
22. Abraham Flexner, "The Usefulness of Useless Knowledge," *Harper's Magazine*, October 1939, 551.

23. Abraham Flexner, *I Remember* (New York: Simon & Schuster, 1940), 366.
24. Louis Bamberger to the Trustees of the Institute for Advanced Study, April 23, 1934, IAS.
25. Willis Lamb, "The Fine Structure of Hydrogen," in *The Birth of Particle Physics*, ed. Laurie M. Brown and Lillian Hoddeson (Cambridge, U.K.: Cambridge University Press, 1983), 322–23.
26. Schwinger to Silvan Schweber, January 15, 1982, in Silvan Schweber, *QED and the Men Who Made It* (Princeton, N.J.: Princeton University Press, 1994), 303.
27. Freeman Dyson to parents, September 25 and October 16, 1947, FJD.
28. Freeman Dyson, "Member of the Club," in *Curious Minds: How a Child Becomes a Scientist*, ed. John Brockman (New York: Pantheon, 2004), 61.
29. Freeman Dyson to parents, October 5, 1941, FJD.
30. Freeman Dyson to parents, March 24, 1943, FJD.
31. Freeman Dyson to parents, May 29, 1943, FJD.
32. Freeman Dyson, interview with Sam Schweber and Christopher Sykes, June 1998, www.webofstories.com/play/freeman.dyson/64.
33. Freeman Dyson, *Selected Papers of Freeman Dyson with Commentary* (Providence: American Mathematical Society, 1996), 10.
34. Ibid., 12.
35. Freeman J. Dyson, "The Electromagnetic Shift of Energy Levels," *Physical Review* 73, no. 6 (1948): 617.
36. Dyson, *Selected Papers*, 12.
37. Dyson, "Electromagnetic Shift of Energy Levels," 617.
38. Freeman Dyson to parents, January 30, 1949, FJD.
39. Minutes of the School of Mathematics, IAS, June 2, 1945, IAS.
40. Freeman Dyson to parents, November 25, 1948, FJD.
41. The Talk of the Town, *New Yorker*, April 30, 1949, 23–24.
42. Verena Huber-Dyson, unpublished memoirs, GBD.
43. Mastrolilli, personal communication, January 17, 2015, GBD.
44. James R. Newman, *The World of Mathematics* (New York: Simon & Schuster, 1956), 1534.
45. Verena Huber-Dyson, unpublished memoirs, GBD.
46. Freeman Dyson interview, Princeton, N.J., May 1, 2004, GBD.
47. Freeman Dyson to parents, October 16, 1948, FJD.
48. Freeman Dyson to Oppenheimer, December 11, 1948, IAS.
49. Freeman Dyson to parents, April 7, 1949, FJD.
50. Freeman Dyson, undated addendum to Kurt Gödel, *The Consistency of the Axiom of Choice and of the Generalized Continuum Hypothesis with the Axioms of Set Theory* (Princeton, N.J.: Princeton University Press, 1940), copy inscribed March 26, 1943, GBD.
51. Penn to Boyle, August 5, 1683, in *Works of Robert Boyle* (London, 1744), 5:646.

4. VOICE OF THE DOLPHINS

1. Leo Szilard, interview recorded in 1963, in *The Collected Works of Leo Szilard: Scientific Papers*, ed. Bernard T. Feld and Gertrud Weiss Szilard (Cambridge, Mass.: MIT Press, 1972), 529.

2. John von Neumann, "Defense in Atomic War" (paper delivered at a symposium in honor of Dr. R. H. Kent, December 7, 1955), *Journal of the American Ordnance Association* (1955): 23.

3. Leo Szilard, "On the Decrease of Entropy in a Thermodynamic System by the Intervention of Intelligent Beings," trans. Anatol Rapoport and Mechthilde Knoller, *Behavioral Science* 9, no. 4 (October 1964): 301.

4. James Clerk Maxwell, *Theory of Heat* (London: Longman's, 1871), 308.

5. Szilard, "On the Decrease of Entropy in a Thermodynamic System by the Intervention of Intelligent Beings," 302.

6. Rutherford to the British Association, September 11, 1933, in A.F., "Atomic Transmutation," *Nature*, September 16, 1933, 433.

7. Szilard, interview recorded in 1963, 529–30.

8. H. G. Wells, "The Discovery of the Future: A Discourse Delivered to the Royal Institution on January 24, 1902," *Nature*, February 6, 1902, 331.

9. H. G. Wells, *The War That Will End War* (London: Frank & Cecil Palmer, 1914).

10. Leo Szilard, *Leo Szilard, His Version of the Facts: Selected Recollections and Correspondence*, ed. Spencer R. Weart and Gertrud Weiss Szilard (Cambridge, Mass.: MIT Press, 1978), 17.

11. S. V. Constant (acting chief of staff, War Department), August 13, 1940, re Enrico Fermi and Leo Szilard, Szilard Papers, UCSD.

12. Manhattan Engineer District, Memorandum re Leo Szilard, December 24, 1946, 1–2, Szilard Papers.

13. Teller to Szilard, July 2, 1945, in *Leo Szilard, His Version of the Facts*, 208.

14. Scharff, interview, La Jolla, Calif., February 8, 1999, GBD.

15. Teller, interview, Palo Alto, Calif., April 22, 1999, GBD.

16. Frederic de Hoffmann, "A Novel Apprenticeship," in *All in Our Time*, ed. Jane Wilson (Chicago: Bulletin of the Atomic Scientists, 1974), 171.

17. Dunne, interview, La Jolla, Calif., July 13, 1998, GBD.

18. Ibid.

19. Theodore B. Taylor, Andrew W. McReynolds, and Freeman John Dyson, "Reactor with Prompt Negative Temperature Coefficient and Fuel Element Therefor," U.S. Patent No. 3,127,325, filed May 9, 1958, issued March 31, 1964.

20. Freeman Dyson to Jeff Bezos, November 22, 2016, GBD.

21. Taylor, interview, August 11, 2000, GBD.

22. Taylor, interview with Gary Marcuse for *The Nature of Things*, Canadian Broadcasting Corporation, 1999.

23. Taylor, interview, Princeton, N.J., May 9, 1998, GBD.

24. Taylor, interview, Princeton, N.J., May 11, 1998, GBD.

25. Ibid.

26. Ibid.

27. Jaromir Astl, interview, Solana Beach, Calif., March 12, 1999, GBD.

28. Taylor, telephone interview, July 2, 1999, GBD.

29. C. J. Everett and S. M. Ulam, "Nuclear Propelled Vehicle, Such as a Rocket," British Patent Specification No. 877,392, application filed February 17, 1960, issued September 13, 1961. Application filed in the United States March 3, 1959.

30. C. J. Everett and S. M. Ulam, "On a Method of Propulsion of Projectiles by Means of External Nuclear Explosions," Los Alamos Scientific Laboratory Report LAMS-1955 (August 1955), 3–5.

31. Taylor, interview, Princeton, N.J., May 9, 1998, GBD.
32. Freeman Dyson, interview, Princeton, N.J., May 11, 1998, GBD.
33. Freeman Dyson to James Lukash, August 15, 1994, GBD.
34. Major Lew Allen (USAF Office of Special Projects) to Ray DeGraff (Air Research & Development Command), May 29, 1958, in "Project Orion: An Air Force Bid for Role in Aerospace," 161–297, in *1964 Annual History of the Air Force Weapons Laboratory, 1 January–December 1964*, comp. Dr. Ward Alan Minge, Captain Harrell Roberts, and Sergeant Thomas L. Suminski, AFSWC, 165.
35. Ibid., 169.
36. S. M. Ulam, testimony, January 22, 1958, before Subcommittee on Outer Space Propulsion of the Joint Committee on Atomic Energy, 85th Cong., 2nd Sess., 44.
37. York, interview, La Jolla, Calif., February 6, 1999, GBD.
38. Ibid.
39. Ibid.
40. Ibid.
41. Air Force Contract AF 18(600)-1812, "Feasibility Study of a Nuclear Bomb Propelled Space Vehicle," June 30, 1958, Exhibit "A"—Statement of Work, 1, AFSWC.
42. Giller and Prickett, interview, Bayfield, Colo., September 15, 1999, GBD.
43. Taylor, interview, Mission Beach, Calif., November 10, 1999, GBD.
44. Freeman Dyson to Verena Huber-Dyson, April 1, 1958, GBD.
45. General Atomic, news release, Washington, D.C., July 2, 1958, GA.
46. Jung to F. J. Dyson, in F. J. Dyson to parents, June 22, 1958, FJD.
47. Freeman Dyson to Oppenheimer, July 4, 1958, IAS.
48. Oppenheimer to Freeman Dyson, July 9, 1958, IAS.
49. Taylor, interview, Princeton, N.J., May 8, 1998, GBD.
50. Taylor, interview, La Jolla, Calif., November 8, 1999, GBD.
51. Freeman Dyson, interview, La Jolla, Calif., July 12, 1998, GBD.
52. F. J. Dyson, *Trips to Satellites of the Outer Planets*, GAMD-490, August 20, 1958, 12, GBD.
53. Mayer, interview, Los Alamos, N.M., September 17, 1999, GBD.
54. Ibid.
55. Theodore B. Taylor, unpublished journal, October 3, 1960, GBD.
56. General Atomic, *Potential Military Applications*, GA-C-962, March 1, 1965, 16–21, AFSWC.
57. John O. Berga, Major, USAF, *Research and Technology Resume, Nuclear Impulse Propulsion Technology Studies, Plan Change*, June 30, 1965, AFSWC.
58. Captain Ronald F. Prater, USAF Project Engineer, *Memorandum for the Record: Sponsorship of ORION Development Planning*, July 22, 1965, AFSWC.
59. F. J. Dyson, "Death of a Project," *Science* 149, no. 3680 (July 9, 1965): 141.
60. F. J. Dyson, "Experiments with Bomb-Propelled Spaceship Models," in *Adventures in Experimental Physics*, Beta issue, ed. Bogdan Maglich (Princeton, N.J.: World Science Education, 1972), 326.
61. Leo Szilard to N. S. Khrushchev, September 30, 1960, 5–6, Szilard Papers.
62. Leo Szilard, "Conversation with Khrushchev on October 5, 1960," in *Toward a Livable World: Leo Szilard and the Crusade for Nuclear Arms Control*, ed. G. Allen Greb, Helen S. Hawkins, and Gertrud Weiss Szilard (Cambridge, Mass.: MIT Press, 1987), 279.
63. Taylor, interview, Princeton, N.J., May 11, 1998, GBD.

64. Leo Szilard, *The Voice of the Dolphins, and Other Stories* (New York: Simon & Schuster, 1961), 22.

65. Murray A. Newman and Patrick L. McGeer, "The Capture and Care of a Killer Whale, *Orcinus orca*, in British Columbia," *Zoologica* 51, no. 2 (Summer 1966): 65.

5. TREE HOUSE

1. Log Salvage Regulations, British Columbia Forest Act.

2. Criminal Code of Canada, Section 339, Offences Resembling Theft.

3. Arthur S. Charlton, *The Belcarra Park Site*, Simon Fraser University Department of Archaeology, Publication No. 9, 1980.

4. Helen Codere, *Fighting with Property: A Study of Kwakiutl Potlatching and Warfare, 1792–1930* (New York: J. J. Augustin, 1950).

5. Charles Hill-Tout, "Later Prehistoric Man in British Columbia," *Transactions of the Royal Society of Canada Section 2* (1895): 103.

6. Simon Fraser, July 2, 1808, in *The Letters and Journals of Simon Fraser, 1806–1808*, ed. W. Kaye Lamb (Toronto: Macmillan, 1960), 106.

7. James Cook, January 30, 1774, in *The Journals of James Cook*, vol. 2, *The Voyage of the Resolution and Adventure, 1772–1775*, ed. James C. Beaglehole (Cambridge, U.K.: Hakluyt Society, 1967), 322.

8. Thomas Edgar, February 13, 1779, Kealakekua Bay, in *The Journals of James Cook*, vol. 3, *The Voyage of the Resolution and Discovery, 1776–1780*, ed. James C. Beaglehole (Cambridge, U.K.: Hakluyt Society, 1967), 1360.

9. George Vancouver, *A Voyage of Discovery to the North Pacific Ocean, and Round the World, in Which the Coast of North-West America Has Been Carefully Examined and Accurately Surveyed* (London: G. G. and J. Robinson, 1798), 1:255.

10. Ibid., 254, 256.

11. Archibald Menzies, June 11, 1792, in *Menzies' Journal of Vancouver's Voyage, April to October, 1792*, ed. C. F. Newcombe (Victoria: Archives of British Columbia, 1923), 53.

12. Menzies, June 23, 1792, in *Menzies' Journal of Vancouver's Voyage*, 60.

13. Vancouver, June 13, 1792, in *Voyage of Discovery*, 1:300.

14. Andrew Paull (Quoitchetahl), in J. S. Matthews, *Early Vancouver* (Vancouver City Archives, 1932), 2:303.

15. Vancouver, June 14, 1792, in *Voyage of Discovery*, 1:302.

16. Police Court—John Hall—Desecration of Indian Graves, before F. G. Claudet and W. J. Armstrong, Esqs., J.P., June 26, 1871, *Mainland Guardian*, June 28, 1871, transcription in Ralph Drew, *Forest and Fjord: The History of Belcarra* (Belcarra, B.C.: Ralph Drew, 2013), 141–42.

17. John Hall's Preemption Claim on Bedwell Bay, Field Notes signed by William G. Pinder, Surveyor, May 23, 1874, Field Book No. 22, Survey and Records Branch, Victoria, B.C.

18. Bell to J. S. Matthews, November 23, 1948, in J. S. Matthews, *Early Vancouver* (Vancouver: City of Vancouver, 2011), 7:61–62.

19. Judge Henry Pering Pelew Crease, *Regina v. John Hall*, New Westminster Assizes "Bench Book" notes, November 27, 1882, British Columbia Archives. Transcription by Michael Cotton, 1998, courtesy of Ralph Drew.

20. Ibid.

21. Ibid.
22. Charles Newland Trew, Coroner, testimony in Crease, *Regina v. John Hall*, November 27, 1882.
23. Crease, in "John Hall Murder Trial," *Mainland Guardian*, December 2, 1882, 3. Transcription courtesy of Ralph Drew.
24. John T. Walbran, *British Columbia Coast Names, 1592–1906* (Ottawa: Government Printing Bureau, 1909), 121.
25. Ethelbert Olaf Stuart Scholefield and Frederic William Howay, *British Columbia, from the Earliest Times to the Present*, vol. 3, *Biographical* (Vancouver: S. J. Clarke, 1914), 609.
26. Decker, in Crease, *Regina v. John Hall*, November 27, 1882.
27. Bole, in "John Hall Murder Trial," *Mainland Guardian*, December 2, 1882, 3.
28. MacElmen, in "John Hall Murder Trial," *Mainland Guardian*, December 2, 1882, 3.
29. Crease, in "John Hall Murder Trial," *Mainland Guardian*, December 2, 1882, 3.
30. Norah Kathleen Bole to W. Briggs Crawford, February 9, 1970, GBD.
31. Malcolm Lowry, *Hear Us O Lord from Heaven Thy Dwelling Place* (New York: Lippincott, 1961), 221.
32. Ibid., 232.
33. George Dyson to Freeman Dyson, December 16, 1972, GBD.
34. John Ambrose Fleming, *The Origin of Mankind: Viewed from the Standpoint of Revelation and Research* (London: Marshall, Morgan & Scott, 1935), 7.

6. STRING THEORY

1. Thor Heyerdahl, *The Kon-Tiki Expedition* (New York: Rand McNally, 1950), 13.
2. Ibid.
3. Clayton Mack, *Bella Coola Man: More Stories of Clayton Mack*, ed. Harvey Thommasen (Madeira Park, B.C.: Harbour, 1994), 180, 184–85.
4. Lydia T. Black, *Russians in Alaska, 1732–1867* (Fairbanks: University of Alaska Press, 2004), 62.
5. Lydia T. Black, "Promyshlenniki . . . Who Were They?," in *Bering and Chirikov: The American Voyages and Their Impact*, ed O. W. Frost (Anchorage: Alaska Historical Society, 1992), 289–90.
6. Catherine II to Denis Chicherin, March 2, 1766, in *Russian Penetration of the North Pacific Ocean, 1700–1797: A Documentary Record*, ed. Basil Dmytryshyn, E. A. P. Crownhart-Vaughan, and Thomas Vaughan (Portland: Oregon Historical Society, 1988), 234–35.
7. Lydia T. Black, "The Russians Were Coming," in *Spain and the North Pacific Coast: Essays in Recognition of the Bicentennial of the Malaspina Expedition, 1791–1792*, ed. Robin Inglis (Vancouver, B.C.: Vancouver Maritime Museum, 1992), 29–34.
8. Martin Sauer, *An Account of a Geographical and Astronomical Expedition to the Northern Parts of Russia* (London: T. Cadell, 1802), 158–59.
9. Ibid., 159.
10. James Cook, October 3, 1778, in *The Journals of James Cook*, vol. 3, *The Voyage of the Resolution and Discovery, 1776–1780*, ed. James C. Beaglehole (Cambridge, U.K.: Hakluyt Society, 1967), 463.
11. Calvin Jay Lensink, "The History and Status of Sea Otters in Alaska" (Ph.D. diss., Purdue University, 1962).

12. Gavriil Ivanovich Davydov, *Two Voyages to Russian America, 1802–1807*, ed. Richard A. Pierce, trans. Colin Bearne (Kingston, Ont.: Limestone Press, 1977), 178.

13. Kyrill T. Khlebnikov, *Colonial Russian America: Kyrill T. Khlebnikov's Reports, 1817–1832*, trans. and ed. Basil Dmytryshyn and E. A. P. Crownhart-Vaughan (Portland: Oregon Historical Society, 1976), 145.

14. Charles Clerke, October 1778, in Beaglehole, *Voyage of the* Resolution *and* Discovery, *1776–1780*, 1334–35.

15. James Cook, October 14 and 15, 1778, in Beaglehole, *Voyage of the* Resolution *and* Discovery, *1776–1780*, 450.

16. Lydia Black, "The Russian Conquest of Kodiak," *Anthropological Papers of the University of Alaska* 24, nos. 1–2 (1992): 170.

17. James Cook, October 19, 1778, in Beaglehole, *Voyage of the* Resolution *and* Discovery, *1776–1780*, 457.

18. Thomas Edgar, October 16, 1778, in Beaglehole, *Voyage of the* Resolution *and* Discovery, *1776–1780*, 1357.

19. Georg Heinrich von Langsdorff, *Voyages and Travels in Various Parts of the World During the Years 1803, 1804, 1805, 1806, and 1807* (London: Henry Colburn, 1813–1814), 2:43.

20. Ivan Veniaminov, *Notes on the Islands of the Unalashka District*, ed. Richard A. Pierce, trans. Lydia T. Black and R. H. Geoghegan (Kingston, Ont.: Limestone Press, 1984), 272.

21. William H. Dall, *Alaska and Its Resources* (Boston: Lee and Shepard, 1870), 241.

22. Halleck, September 22, 1868, in *Annual Report of the Secretary of War for the Year 1868* (Washington, D.C.: Government Printing Office, 1868), 37–40.

23. William S. Laughlin, John D. Heath, and Eugene Arima, "Two Nikolski Aleut kayaks: iqyax̂ and ulux̂tax̂ from Umnak Is.," in *Contributions to Kayak Studies*, ed. Eugene Arima (Hull, Quebec: Canadian Museum of Civilization, 1991), 163–209.

24. Laughlin, personal communication, April 5, 1991. See William S. Laughlin, S. B. Laughlin, and S. B. Beman, "Aleut Kayak-Hunter's Hypertrophic Humerus," *Current Research in the Pleistocene* 8 (1991): 55–56.

25. George Vancouver, July 13, 1792, in *A Voyage of Discovery to the North Pacific Ocean, and Round the World, in Which the Coast of North-West America Has Been Carefully Examined and Accurately Surveyed* (London: G. G. and J. Robinson, 1798), 1:336.

26. *Extract of the Diary of Galiano and Valdés*, July 13, 1792, in *Spanish Explorations in the Strait of Juan de Fuca*, ed. Henry R. Wagner (Santa Ana, Calif.: Fine Arts Press, 1933), 217.

27. *Voyage of the* Sutil *and* Mexicana, in Wagner, *Spanish Explorations in the Strait of Juan de Fuca*, 275–78.

28. George Dyson to family, August 10, 1973, GBD.

29. Baranov, quoted in Kiril T. Khlebnikov, *Baranov: Chief Manager of the Russian Colonies in America*, ed. Richard A. Pierce, trans. Colin Bearne (Kingston, Ont.: Limestone Press, 1973), 44–45.

30. Carl Woese, "A New Biology for a New Century," *Microbiology and Molecular Biology Reviews* 68, no. 2 (June 2004): 176.

31. Thor Heyerdahl, *American Indians in the Pacific* (London: George Allen and Unwin, 1952), 158.

7. EREWHON REVISITED

1. Butler to the Reverend Charles Seager, May 24, 1833, Shropshire Archives, Shrewsbury, U.K.

2. Charles Darwin (1876), *Autobiography*, in *Life and Letters of Charles Darwin*, ed. Francis Darwin (New York: Appleton, 1896), 1:29.

3. Thomas Butler, in Darwin, *Life and Letters of Charles Darwin*, 1:144.

4. George Bernard Shaw, *Back to Methuselah: A Metabiological Pentateuch* (London: Constable, 1921), xliii.

5. Thomas Butler to Samuel Butler, August 3, 1859, in *The Family Letters of Samuel Butler*, ed. Arnold Silver (Stanford, Calif.: Stanford University Press, 1962), 89.

6. Samuel Butler, *A First Year in Canterbury Settlement* (London: Longman & Green, 1863), in *Canterbury Settlement and Other Early Essays*, vol. 1 of *The Shrewsbury Edition of the Works of Samuel Butler*, ed. Henry Festing Jones (London: Jonathan Cape, 1923), 82.

7. Ibid., 65.

8. Ibid., 72–73.

9. Ibid., 82.

10. Ibid., 87.

11. Williams to Henry Festing Jones, August 19, 1912, in *Samuel Butler: A Memoir*, ed. Henry Festing Jones (London: Macmillan, 1919), 1:84.

12. Robert B. Booth, *Five Years in New Zealand* (London: privately printed, 1912), 73.

13. Ellen Shephard Tripp, *My Early Days* (Canterbury, N.Z.: Whitcombe and Tombs, 1916), 25.

14. Ibid., 8.

15. James McNeish, "The Man Between the Rivers," *New Zealand Geographic*, October–December 1990.

16. Butler to Darwin, October 1, 1865, in Jones, *Canterbury Settlement and Other Early Essays*, 186.

17. Samuel Butler, *Unconscious Memory* (London: David Bogue, 1880), vol. 6 of *The Shrewsbury Edition of the Works of Samuel Butler* (London: Jonathan Cape, 1924), 12.

18. Darwin to an unidentified editor, March 24, 1863, in Jones, *Canterbury Settlement and Other Early Essays*, 184–85.

19. Samuel Butler, "Barrel-Organs," *Press* (Christchurch), January 17, 1863, reprinted in Jones, *Canterbury Settlement and Other Early Essays*, 196.

20. Samuel Butler, "Darwin Among the Machines," *Press* (Christchurch), June 13, 1863.

21. E. C. Richards, ed., *Diary of E. R. Chudleigh, 1862–1921* (Christchurch, 1950), 68.

22. Samuel Butler, "The Mechanical Creation," *Reasoner* (London), July 1, 1865, reprinted in Jones, *Canterbury Settlement and Other Early Essays*, 231–33.

23. Samuel Butler, note, June 1887, in Jones, *Samuel Butler: A Memoir*, 1:155.

24. Samuel Butler, 1901, in Jones, *Samuel Butler: A Memoir*, 1:158.

25. Jones, *Samuel Butler: A Memoir*, 1:155.

26. Butler to Darwin, May 11, 1872, in Jones, *Samuel Butler: A Memoir*, 1:156–57.

27. Butler to Darwin, May 30, 1872, in Jones, *Samuel Butler: A Memoir*, 1:158.

28. Samuel Butler, *The Fair Haven: A Work in Defence of the Miraculous Element in Our Lord's Ministry upon Earth, Both as Against Rationalistic Impugners and Certain Orthodox Defenders, by the Late John Pickard Owen, Edited by William Bickersteth Owen, with a Memoir of the Author* (London: Trübner, 1873).

29. John Butler Yeats, "Recollections of Samuel Butler," in *Essays, Irish and American* (Dublin: Talbot Press, 1918), 17.

30. Samuel Butler, *Evolution, Old and New; or, The Theories of Buffon, Dr. Erasmus Darwin, and Lamarck, as Compared with That of Charles Darwin* (London: Hardwicke & Bogue, 1879).

31. Darwin to Huxley, February 4, 1880, in Jones, *Samuel Butler: A Memoir*, 2:454.

32. Huxley to Darwin, February 3, 1880, Cambridge University Library.

33. *Saturday Review* (London), May 31, 1879, 682.

34. Erasmus Darwin, *Zoonomia; or, The Laws of Organic Life* (London: J. Johnson, 1794), 1:509.

35. Samuel Butler, *Luck, or Cunning, as the Main Means of Organic Modification? An Attempt to Throw Additional Light upon Darwin's Theory of Natural Selection* (London: Trübner, 1887), vol. 8 of *The Shrewsbury Edition of the Works of Samuel Butler* (London: Jonathan Cape, 1924), 61.

36. Ibid., 234.

37. Samuel Butler, *Life and Habit* (London: Trübner, 1878), 128–29.

38. Butler, *Unconscious Memory*, 57.

39. Ibid., 16.

40. Samuel Butler, *Erewhon; or, Over the Range* (London: Trübner, 1872), 189.

41. Ibid., 203.

42. Samuel Butler, "From our Mad Correspondent," *Press* (Christchurch), September 15, 1863, reprinted in Joseph Jones, *The Cradle of "Erewhon": Samuel Butler in New Zealand* (Austin: University of Texas Press, 1959), 196–97.

43. Esther Dyson to Verena Huber-Dyson, October 24, 1967, GBD.

44. John von Neumann, *Lectures on Probabilistic Logics and the Synthesis of Reliable Organisms from Unreliable Components, from Notes by R. S. Pierce of Lectures at the California Institute of Technology, January 4–15, 1952*, in *Automata Studies* (Princeton, N.J.: Princeton University Press, 1956), 43–99.

45. George Dyson, 1980, GBD.

46. Stewart Brand, *Whole Earth Software Catalog* (Garden City, N.Y.: Doubleday, 1984), 13.

47. Julian H. Bigelow, "Theories of Memory," in *Science in the Sixties: The Tenth Anniversary AFOSR Scientific Seminar, Cloudcroft, New Mexico, June 1965*, ed. David L. Arm (Albuquerque: University of New Mexico Press), 86.

48. Ibid., 85–86.

49. Julian Bigelow, "Computer Development at the Institute for Advanced Study," in *A History of Computing in the Twentieth Century*, ed. Nicholas Metropolis, J. Howlett, and Gian-Carlo Rota (New York: Academic Press, 1980), 304.

50. Pomerene, interview with Nancy Stern, September 26, 1980, oral history 31, CBI.

51. Samuel Butler, "Material for *Erewhon Revisited*," in *The Note-Books of Samuel Butler*, ed. Henry Festing Jones (London: Fifield, 1912), 289.

52. Butler, *Erewhon*, 188–89.

53. Ibid., 198–99.

54. Ibid., 216–17.

55. Stanislaw Ulam, in *Mathematical Challenges to the Neo-Darwinian Interpretation of Evolution: A Symposium Held at the Wistar Institute, April 25–26, 1966*, ed. Paul S. Moorhead and Martin M. Kaplan (Philadelphia: Wistar Institute, 1966), 42.

56. Butler, *Erewhon*, 197.

8. NO TIME IS THERE

1. G. W. Leibniz, *Discourse on the Natural Theology of the Chinese*, ed. and trans. Henry Rosemont Jr. and Daniel J. Cook (Honolulu: University of Hawaii Press, 1977), 158.

2. G. W. Leibniz, "De Progressione Dyadica—Pars I" (MS. March 15, 1679), published in facsimile (with German translation) in *Herrn von Leibniz' Rechnung mit Null und Eins*, ed. Erich Hochstetter and Hermann-Josef Greve (Berlin: Siemens Aktiengesellschaft, 1966), 46–47. English translation by Verena Huber-Dyson, 1995.

3. Jack Rosenberg, "The Computer Project," unpublished draft, February 2, 2002, GBD.

4. Slutz, interview with Christopher Evans, June 1976, oral history 086, CBI.

5. J. H. Bigelow et al., *Interim Progress Report on the Physical Realization of an Electronic Computing Instrument*, January 1, 1947, 15–16, IAS.

6. Pomerene, interview with Nancy Stern, September 26, 1980, oral history 31, CBI.

7. Bigelow et al., *Interim Progress Report*, January 1, 1947, 54.

8. Julian Bigelow, "Computer Development at the Institute for Advanced Study," in *A History of Computing in the Twentieth Century*, ed. Nicholas Metropolis, J. Howlett, and Gian-Carlo Rota (New York: Academic Press, 1980), 297.

9. Institute for Advanced Study Electronic Computer Project, *Monthly Progress Report: July and August, 1947*, 2–3, IAS.

10. J. H. Bigelow et al., *Fifth Interim Progress Report on the Physical Realization of an Electronic Computing Instrument*, January 1, 1949, 31, IAS.

11. Bigelow, interview with Flo Conway and Jim Siegelman, October 30, 1999, courtesy of Flo Conway and Jim Siegelman.

12. John Aubrey, *Aubrey's Brief Lives: Edited from the Original Manuscripts by Oliver Lawson Dick* (Ann Arbor: University of Michigan Press, 1949), 167.

13. Hooke, May 7, 1673, in R. T. Gunther, *Early Science in Oxford* (Oxford: printed for the author, 1930), 7:412.

14. Aubrey, *Aubrey's Brief Lives*, 167.

15. Robert Hooke (1682), *The Posthumous Works of Robert Hooke, Containing His Cutlerian Lectures, and Other Discourses, Read at the Meetings of the Illustrious Royal Society*, ed. Richard Waller (London: Richard Waller, 1705), 140.

16. Ibid.

17. Ibid., 144.

18. Ibid., 134.

19. *The Famous History of Frier Bacon, Containing the Wonderful Things That He Did in His Life; Also the Manner of His Death, with the Lives and Deaths of the Two Conjurers Bungey and Vandermast. Very Pleasant and Delightful to Be Read* (London: T. Passenger, 1679), 12–13.

20. Ibid., 15.

21. Ibid., 17.

22. Orders of Major General J. M. Schofield, issued under Lieutenant General Sheridan, General Orders No. 8, November 20, 1871, in *Report of the Commissioner of Indian Affairs for the Year 1871* (Washington, D.C.: Government Printing Office, 1872), 94.

23. Ibid.

24. John Gregory Bourke, *On the Border with Crook* (New York: Scribner's, 1891), 213.

25. Orders of Schofield, issued under Sheridan, General Orders No. 8, November 20, 1871, 95.

26. George Crook, General Orders No. 14, April 9, 1873.

27. People's Republic of China, State Council, "Planning Outline for the Construction of a Social Credit System (2014–2020)," June 14, 2014, accessed October 7, 2018, as translated at www.chinalawtranslate.com/socialcreditsystem.

28. Von Neumann to Teller, October 7, 1946, Von Neumann Papers, Library of Congress.

9. CONTINUUM HYPOTHESIS

1. William Tecumseh Sherman, *Memoirs of General William T. Sherman* (New York: Appleton, 1889), 2:413.

2. Ibid., 1:11.

3. Jefferson to John Adams, April 20, 1812, National Archives, accessed April 11, 2019, founders.archives.gov/documents/Jefferson/03-04-02-05_7.

4. Sherman, *Memoirs*, 1:19–20.

5. Sherman to Sheridan, October 7, 1872, microfilm reel no. 17, Sheridan Papers, in David D. Smits, "The Frontier Army and the Destruction of the Buffalo, 1865–1883," *Western Historical Quarterly* 25, no. 3 (Autumn 1994): 335.

6. Sherman, *Memoirs*, 2:413–14.

7. Ibid., 436.

8. William Temple Hornaday, *The Extermination of the American Bison*, in *Annual Report of the Smithsonian Institution* (Washington, D.C., 1889), 529.

9. Ibid., 530.

10. Ibid., 525, 545.

11. Schonchin, statement, ca. 1932, in Cora Du Bois, *The 1870 Ghost Dance* (Berkeley: University of California Press, 1939), 10.

12. McLaughlin, October 17, 1890, in James Mooney, *The Ghost-Dance Religion and the Sioux Outbreak of 1890*, in *Fourteenth Annual Report of the Bureau of Ethnology, Part 2* (Washington, D.C.: Government Printing Office, 1896), 787.

13. Mooney, *Ghost-Dance Religion and the Sioux Outbreak of 1890*, 657.

14. Ibid., 870.

15. John W. Comfort (1892), in "Of Memory and Massacre: A Soldier's Firsthand Account of the 'Affair on Wounded Knee,'" ed. Karl Jacoby, *Princeton University Library Chronicle* 64, no. 2 (Winter 2003): 345–46.

16. Ibid., 348–51.

17. Donald F. Danker, ed., "The Wounded Knee Interviews of Eli S. Ricker," *Nebraska History* 62, no. 2 (Summer 1981): 176–78.

18. Mooney, *Ghost-Dance Religion and the Sioux Outbreak of 1890*, 878–79.

19. Ibid., 765.

20. Warner to Mooney, October 12, 1891, in ibid., 767.

21. Mooney, *Ghost-Dance Religion and the Sioux Outbreak of 1890*, 768.

22. Ibid., 770.

23. Ibid., 772–73.

24. Ibid., 775–76.

25. Julian Bigelow, Arturo Rosenblueth, and Norbert Wiener, "Behavior, Purpose, and Teleology," *Philosophy of Science* 10, no. 1 (1943): 18–24.

26. Norbert Wiener, *God and Golem, Inc.* (Cambridge, Mass.: MIT Press, 1964), 69.

27. Ulam, quoted by Gian-Carlo Rota, "The Barrier of Meaning," *Letters in Mathematical Physics* 10 (1985): 99.

28. Freeman Dyson to parents, March 29, 1943, FJD.

29. Kurt Gödel, "What Is Cantor's Continuum Problem?," *American Mathematical Monthly* 54 (1947): 516.

30. Von Neumann to Oswald Veblen, November 30, 1945, Veblen Papers, Library of Congress.

31. David Hilbert, "Mathematical Problems, Lecture Delivered Before the International Congress of Mathematicians at Paris in 1900," *Bulletin of the American Mathematical Society* 8 (July 1902): 446.

32. Kurt Gödel, *The Consistency of the Axiom of Choice and of the Generalized Continuum Hypothesis with the Axioms of Set Theory* (Princeton, N.J.: Princeton University Press, 1940), 35.

33. Leibniz to Simon Foucher, *Journal des Savans*, March 16, 1693, in *The Philosophical Works of Leibnitz*, trans. George Martin Duncan (New Haven, Conn.: Tuttle, Morehouse & Taylor, 1890), 65.

ACKNOWLEDGMENTS

Children are like cats: people either like them or don't like them. Both cats and children learn to tell the difference at an early age. I remember being taken to a cocktail party in Princeton, New Jersey, with my mother and father where I was left alone in a back room in a crib. I was unhappy with my captivity until a rotund, friendly gentleman, who might have been "Johnny" von Neumann, abandoned his colleagues, talked to me, and gave me a sip of his drink through the bars of the crib. Von Neumann, one of the architects of modern digital computing, held only one patent: for a method of non–von Neumann computing, the subject of this book.

Nonconformists are also like cats, making their way through life the way alley cats survive thanks to people who leave out food. Jim Bates, Michael Berry, Jeff Bezos, Julian Bigelow, Lydia Black, Béla and Gabriella Bollobás, John Brockman, Anne Hus Brower, Barbara Brower, David R. Brower, John Brower, Kenneth Brower, Robert Irish Brower, Kathy

View of Mount Edgcumbe *when the* Cape *bears N.º W. 4 leaª disª*

Cain, Eric Chinski, Ry and Susie Cooder, Frederica de Laguna, Esther Dyson, Imme Dyson, Lauren Dyson, Richard Elson, Jason Halm, Katarina Halm, Michael J. Hawley, Danny Hillis, Robert Hunter, George F. Jewett, Huey D. Johnson, Charles Jurasz, Ellen Lackerman, William S. Laughlin, Will Malloff, Katinka Matson, Charles McAleese, Stefan McGrath, James Noyes, Richard P. O'Neill, Tim O'Reilly, Richard A. Pierce, Charles Simonyi, Paul Spong, Ulli Steltzer, Neal Stephenson, Greg Streveler, Jin Tatsumura, Theodore B. Taylor, Peter Thomas, Françoise Ulam, Carl Woese, Ann Yow, and Joe Ziner, in alphabetical order, left food out for me along the way, as did the archivists and librarians who granted access to their collections as if I had all the credentials in the world.

The Canada Council, Friends of the Earth, Fairhaven College of Western Washington University, the Institute for Advanced Study, Blue Origin LLC, and the University of Victoria lent institutional support.

My father, Freeman Dyson, who died as this book was going to press, climbed trees, walked fast, and dragged me out of bed in October 1957 to see *Sputnik* fly overhead. When I showed him a copy of an old General Atomic Calculation Sheet from 1959 titled *Outer Planet Satellites*, comparing possible destinations for the voyage he was planning at the time, he studied it for a moment and said, "Enceladus still looks good."

My mother, Verena Huber-Dyson, who died in 2016, had no tolerance for terms like "quantum," "undecidable," and "singularity," with precise mathematical meaning, being applied to fuzzy nonmathematical ideas. She would have objected to "Continuum Hypothesis" being used as a chapter title in this book. "This will all go smoothly. Let's get going," were her last words.

INDEX

Admiralty Point, 142, 157, 161
Adney, Edwin Tappan, 183
Advanced Research Projects Agency
    (ARPA), 116–18, 123–24
advertising, 218
Afghanistan, 62
aircraft, 88, 89, 249–50
Air Force, U.S. (USAF), 116, 118, 124;
    Special Weapons Center (AFSWC),
    124–26
airline travel, 250
Alamogordo (Trinity) test, 109, 118
Alaska, 8, 148, 179–81
Alaska Native Claims Settlement Act, 180
Albion Stove Works, 152
Aleutian Islands, 23–24, 171–74, 178, 181
Aleuts (Unangan), 23–24, 32, 37, 38, 172,
    180–81; kayaks of, see kayaks, Aleut
algorithms, 7, 10, 245, 250, 251
Ali, Muhammad, 130
Alope, 55–56
alphabet, 38, 39
Alutiiq, 176
American de Forest Wireless Telegraph
    Company, 79–80
American Indians in the Pacific (Heyerdahl),
    195
American Marconi Company, 82
American Physical Society, 86, 91
amino acids, 227

analog computing, 6, 8, 9, 245, 248–49, 252;
    continuum hypothesis and, 10, 246–47,
    249; digital computing vs., 5, 6, 8, 106,
    133–34, 247–48
analog universe, time in, 10, 227, 228
analog waveforms, 81
Anangula, 23
Anglo-American Telegraph Company, 76
Anthropocene, 103
Apaches, 8–10, 24, 39–69, 175, 221, 232, 239;
    Chiricahua, 8–10, 45–47, 49, 54, 55,
    66–67, 233; exile and captivity of, 65–69;
    name of, 43–44; numbered identity tags
    for, 10, 52, 54, 232–33; on San Carlos
    reservation, 51–52, 54, 56–57, 66, 67; tribal
    and group divisions among, 43; White
    Mountain, 40, 54
aphids, humans as, 219
Apollo moon landing, 125, 131
Apple, 211, 214–15, 234
Army, U.S., 9, 40, 51–53, 58, 61, 108, 110–11,
    130, 238, 239
Army Signal Corps, U.S., 62–63, 221
ARPA (Advanced Research Projects
    Agency), 116–18, 123–24
Arran Rapids, 185
artificial intelligence (machine intelligence),
    4, 6, 7, 10, 208, 217–19, 245, 251–53; brass
    head and, 230–32; three laws of, 251–52;
    two approaches to, 222

Ashby, W. Ross, 251
Asia-Americas connection, 3, 8, 23, 143
Astl, Jaromir, 114
Astor, John Jacob, 156
Athenaeum, 205
Atkey, Freeman, 88
atomic bomb, *see* nuclear weapons
atomic energy, 107, 108
Atomic Energy Commission (AEC), 91, 92, 113
Atoms for Peace conference, 111
Aubrey, John, 228, 229
Audion, 80, 84
automata, 212, 217–18, 226, 252
Avacha, 25–26, 30

Babbage, Charles, 61
Back to Methuselah (Shaw), 200
Bacon, Roger, 230–31
Bad Pyrmont, 3, 15–16, 224
Baer, Reinhold and Marianne, 97
baidarkas, *see* kayaks, Aleut
Baird, George W., 42
Baird, Spencer F., 239
Bamberger, Louis, 83, 84, 100
Baranov, Alexander, 190
Bark Canoes and Skin Boats of North America
    (Adney and Chapelle), 183–84
Barksdale, James, 211
Bartlett Cove, 190, 191
Basov, Emel'ian, 171
Bates, Jim, 131–32
Baylor, John R., 48–49
beachcombers, 140–41
Beagle, HMS, 9
"Behavior, Purpose, and Teleology"
    (Bigelow, Rosenblueth, and Wiener), 245
Belcarra, 144, 147–51, 155–56, 158, 159, 211
Belcarra Regional Park, 142, 147
Bell, J. Warren, 151
Bella Coola, 170
Bell Telephone Laboratories, 82, 102
Berga, John O., 126
Bering, Vitus, 4, 8, 10–11, 17–20, 24, 26, 171, 174, 190, 191; death of, 27, 28
Bering-Chirikov expedition, *see* Great Northern Expedition

Beringia (land bridge), 23, 29, 143, 240
Bering Island, 20, 28–30, 32, 171
Bering Sea, 20, 23
Bering Strait, 1, 20
Bernoulli, Johann, 16
Berry, Michael, 158–59
Bethe, Hans, 84, 86–87, 89–90, 112, 114
Betty L., 192
Betzinez, Jason, 57
Bezezekoff, Steve, 181
Bigelow, Julian, 215–17, 224–28, 245
Big Foot, 242
Bikini (Crossroads) test, 118
Billings, Joseph, 174, 177
binary arithmetic, 4–6, 9, 223, 234
bioluminescence, 193, 194
biometric identity tags, 234
Birch Bay, 146–47
Bird Island, 21
bison (buffalo), 60, 237–39, 241
bits, 94, 106, 222, 227–28, 248, 249, 252; in shift registers, 223–26, 228
Black, Lydia, 172
Black Coyote, 244
Blackfish Sound, 132–33
Bland family, 182
Bletchley Park, 89
boats, 138, 169; design optimization of, 175, 181–82, 193–94; drifting logs and, 139–40; Peter the Great and, 15; *shitiki*, 172; *see also* dugout canoes; kayaks; skin boats
Bohr, Niels, 112, 125
Bole, Florence Coulthard, 156
Bole, Norah Kathleen, 156, 157
Bole, Norman, 153–56
Bole, Percy Hampton, 155–57
Booth, Robert, 202
Bourke, John Gregory, 52, 57–59, 233
Boyle, Robert, 100
brain, 6, 7, 134, 229, 249
Brand, Stewart, 213
Brant, Karsten, 15
brass head, 230–32
British Columbia, 9, 131, 137–38, 148–49, 168, 179, 183; logging in, 137–40
Brodsky, Dina, 181
Broughton, William, 144, 146, 184–85
Brower, Barbara, 130

Brower, David, 130
Brower, Kenneth, 191
buffalo, 60, 237–39, 241
buffers, 222
Buffon, Georges, 206
Bühler AG, 94
Bulletin of the Atomic Scientists, 110
Bungey, Frier, 230–31
Bureau of American Ethnology, 241
Bureau of Indian Affairs, 51
Burmah, 201
Burrard Inlet, 9, 142, 144, 147–50, 152, 156–57, 164, 184
Butler, Samuel (headmaster and bishop), 199–200, 208
Butler, Samuel (novelist), 200–210, 212, 219; "Darwin Among the Machines," 10, 204; *Erewhon*, 10, 200, 205, 208–10, 212, 217, 219; *Erewhon Revisited*, 10, 217–18; *The Evidence for the Resurrection of Jesus Christ*, 206; *Evolution, Old and New*, 207; *The Fair Haven*, 206; *Life and Habit*, 205; "The Mechanical Creation," 204–205; *The Way of All Flesh*, 200
Butler, Thomas, 200

Cabeza de Vaca, Álvar Núñez, 44–45
calculus, 5, 7, 228
calculus ratiocinator, 4, 228
California, 148, 179
Calvert Island, 195, 196
Cambridge University, 73, 87, 200; Trinity College at, 88, 89
Canada, 131, 140–41, 148, 156
Canal de Floridablanca, 145
canoes, dugout, *see* dugout canoes
Canterbury Settlement, 200–201
Cantor, Georg, 246–48
Carleton, James H., 49
Carson, Kit, 49
Castañeda de Nájera, Pedro de, 45
Castillo Maldonado, Alonso del, 44
Castle Bravo test, 92
Catherine I of Russia, 17
Catherine II of Russia (Catherine the Great), 173–75, 178
cathode rays, 80–82, 101–102, 216
Caulder, Peter, 151–52, 154

Cavalry, U.S., 40, 63, 130, 175
Cavendish Laboratory, 73
cedar, 137, 160, 164–65, 185; shakes, 159–60
cellular automata, 226
Century Magazine, 42
chain reactions, 107, 227
Chambers, Robert, 206
Chapelle, Howard, 183–84
Chatham, 144, 146, 185
Chealtah, 151, 152, 154
chemistry, 96–97
Chichagof Island, 33–35
Chicherin, Denis, 173
Chichilticale, 45–47
Chihuahua, 58
Chihuahua, Eugene, 67, 68
China, 4, 5, 172, 223, 234
Chiricahua Apaches, 8–10, 45–47, 49, 54, 55, 66–67, 233; *see also* Apaches
Chirikov, Alexei, 4, 8, 10–11, 17, 19, 32–36, 165, 174, 190–92
Christchurch, 202–203, 209
Christchurch-Lyttelton telegraph line, 209–10
Christchurch *Press*, 203–204, 209, 212
Christianity, 73, 200, 240
Christie, Julie, 158
chronometers, 229
Chudleigh, Edward, 204
Chugach, 171, 176
Church of St. Nicholas, 171
Civil War, 48–51, 60, 61, 180, 237, 238, 245
Clerke, Charles, 145, 177
Cleveland, Grover, 58–60, 66, 67
clocks, 227–28
Clovis tradition, 167
Clyde, Norman, 130
Cochise, 49, 55, 56
code-breaking computers, 224
codes and coding, 4–7, 9, 39, 61, 63, 71, 94, 134–35, 208, 213, 216, 222, 226–28, 248–49, 250, 252
Cohen, Paul, 247
Columbia University, 85
Comanches, 24
Comfort, John W., 242–43
communication, 133–35, 221–22; machine readability and, 222

complex systems, defined, 251
computational fluid dynamics, 92, 193
computers: Apple, 211, 214–15, 234;
  code-breaking, 224; memory for, 216–17;
  personal, 211–16; software for, 213;
  see also digital computing
Conner, Daniel Ellis, 49–50
consciousness, 37–38, 134, 209, 212, 218
Consistency of the Continuum Hypothesis,
  The (Gödel), 99, 246
continuum, 10, 227, 246–48, 252, 253
continuum hypothesis, 10, 99, 246–49
control, 6–8, 10, 208, 218, 245, 249–53
Cook, James, 124, 145, 146, 165, 174–75,
  177–78
Cordero Channel, 185
Cornell University, 87
Coronado, Francisco Vázquez de, 45–47
Cotter, Rick, 158
Crease, Henry P. P., 153, 155
Crimean War, 179, 180
Crook, George, 52–54, 56–60, 63, 66, 68,
  232–33, 238
Crossroads (Bikini) test, 118
Cruse, Thomas, 41
cybernetics, 245

Daily Telegraph, 77
Daklugie, Asa "Ace," 41, 51–52, 60, 68, 69
Dall, William H., 180
Daly, Henry, 64
DARPA, 116
Darwin, Charles, 9, 10, 73, 200, 203–209; On
  the Origin of Species, 200, 201, 203–204,
  206
Darwin, Erasmus, 204, 206, 207
"Darwin Among the Machines" (Butler), 10,
  204
data centers, 233–34; Apple, 234; National
  Cybersecurity Initiative Data Center,
  221–22, 228, 230, 233
Dauenhauer, Nora and Richard, 35–36
Davis, Britton, 40–41, 51, 52, 54–55
Davis, Jefferson, 48–49, 180
Davydov, Gavriil, 176
deadheads, 139–40
Decker, Stephen, 150–52, 154
Deep Cove, 159

Deep Space Force, 124–25
de Forest, Lee, 78–80, 84, 218, 225
de Hoffmann, Frederic "Freddy," 110, 111,
  116, 118–19
Dement'ev, Avraam, 32–33, 35, 36
Dent Island, Dent Rapids, 185
de Remond, Nicolas, 223
Design for a Brain (Ashby), 251
Design of Computers, Theory of Automata, and
  Numerical Analysis (von Neumann), 212
Desolation Sound, 184
Devil's Hole, 185
Dietz, George, 149
digital computing, 4–8, 10, 92, 94, 105,
  106, 165, 208, 212–17, 223–35, 245–49,
  252; analog computing vs., 5, 6, 8, 106,
  133–34, 247–49; continuum hypothesis
  and, 10, 246–47, 249; noise and, 248;
  self-replication and, 7, 208, 227; structure
  and sequence in, 222; time and, 10; see also
  bits; computers
digital information networks, 9, 39, 61, 69, 71,
  211, 214, 222, 233–34, 249–50
digital revolution, 4, 7, 9, 61, 252
digital universe, 7, 8, 10, 222, 252; time in,
  227–28
Dineh, 43
Dirac, Paul, 85–86, 88
Discourse on the Natural Theology of the Chinese
  (Leibniz), 223
Discovery, 144–47, 174, 178, 185
Discovery Passage, 184–85
Dixon Entrance, 137
DNA, 6, 227, 234, 248
Dobroe Namerenie, 174
Dollarton, 157
dolphins: killer whales, 132–35; Voice of
  the Dolphins, The (Szilard), 9, 127–29,
  134–35
Dorantes, Andrés, 44
drift logs, 137, 139–42
D'Sonoqua, 131–33, 141–42, 144, 158–59, 161,
  182–85
dugout canoes, 160, 176, 183, 194–95, 222;
  skin boats vs., 170–71
Dukas, Helen, 168, 170
Dunne, Brian, 111, 124
Dyson, Esther, 99–100, 210–11, 213

Dyson, Freeman, 87–91, 94, 97–98, 111, 121, 246; Feynman and, 90–91; letter to Oppenheimer from, 119–20; Orion and, 116, 117, 119–26; space exploration as interest of, 116–17; Taylor and, 114, 117, 120, 121; TRIGA reactor and, 111–12; Verena and, 98–99, 119

Dyson, George: childhood and teen years of, 99–102, 119–21, 126–27, 168–69, 212; D'Sonoqua and, 131–33, 141–42, 144, 158–59, 161, 182–85; in high school, 129–31; kayaking and, 129, 131, 169, 182, 183, 185–86, 191–93, 195; mountaineering and, 129–30; tree house of, 161–64, 169, 184, 210

Dyson, George (grandfather of author), 88

Dyson, Lauren, 219

Dyson, Mildred Atkey, 88

East India Company, 15, 175

eddies and whirlpools, 185–86, 193–95

Eddington, Arthur, 88

Edgar, Thomas, 145–46, 178

Edison, Thomas, 71–73, 82

Edison effect, 71–72, 74, 77, 85

Edison Electric Light Company, 72–74, 78

Einführung in die Quantentheorie der Wellenfelder (Wentzel), 90

Einstein, Albert, 84, 91, 92, 102, 105, 108, 110, 168; compass given to, 169; Dukas and, 168; Szilard and, 106, 107

Einstein, Margot, 168

Eisenhower, Dwight D., 116, 117

electromagnetic fields, 169

electronic, use of term, 73

Electronic Computer Project (ECP), *see* Institute for Advanced Study

electronics, 5–7, 9, 71, 81–82

electrons, 9, 10, 71–74, 78, 80–82, 88, 102, 224, 225, 249; cathode rays, 80–82, 101–102, 216

Eliot, T. S., 92

Emergency Committee in Aid of Displaced German Scholars, 91

Emergency Committee of Atomic Scientists, 92, 110

Enceladus, 122, 123

Eniwetok, 113

entropy, and information, 105–106

epochs of technology, 7–8, 135, 252

Erewhon; or, Over the Range (Butler), 10, 200, 205, 208–10, 212, 217, 219

Erewhon Revisited (Butler), 10, 217–18

Ermak, 190

Esteban (Estevanico) de Dorantes, 44–47

Evelyn, John, 15

Everett, Cornelius J., 115

Evidence for the Resurrection of Jesus Christ, as Given by the Four Evangelists, Critically Examined, The (Butler), 206

evolution, 73, 205–209, 219, 223; intelligence and, 207, 208; machines and, 208–209, 217–19

Evolution, Old and New (Butler), 207

Fair Haven, The (Butler), 206

FBI, 107–109

Federal Express, 211

Fermi, Enrico, 84, 86, 108

Feynman, Richard, 84, 86, 90–91

field theory, 90, 98

Fireweed, 133

First Kamchatka Expedition, 17

fjords, 142

Fleming, John Ambrose, 71–78, 164, 218, 225; de Forest and, 78, 218

Fleming valves, 78–80

Flexner, Abraham, 83–84

flip-flops (toggles), 225

Flood-Page, Major, 75

Flowers, Thomas, 224

Fly, Camillus S., 58

flying squirrels, 163

Forbes, 210–11

Fort Langley, 148

Fort Ross, 179

Fort Rupert, 148

Fowler, Orson S., 50

Frankenstein (Shelley), 232

Fraser, Simon, 144

Fraser River, 142, 144, 147–49

Free, Mickey, 53

French, Bruce, 86

Fuld, Carrie Bamberger, 83, 84, 100

functions, continuous vs. discrete, 5, 9, 248, 250, 252

fur trade, 172–73, 176–79

INDEX

Gadsden Purchase, 48
Galiano, Dionisio Alcalá, 144, 148, 185
Galois, Évariste, 96
Gatewood, Charles, 54, 64–65
Gauss, Carl Friedrich, 61
General Atomic (GA), 110–12, 114–16, 118–20, 122–27
General Dynamics, 110, 125
General Motors 6–71 engine, 186–87
genes, 6, 208, 227
George, Chief Dan, 149
Geronimo (Goyaałé), 39–44, 49, 53–60, 62–68, 71; surrender of, 65–67
Ghost Dance movement, 10, 240–41, 243–44
Giller, Ed, 118
Glacier Bay, 189–91
Glacier Bay National Monument, 190
Gmelin, Johann Georg, 17, 19
Gödel, Kurt, 99, 212, 234, 246, 247
gold, 48, 148, 149, 179
Goldstine, Herman, 224
golem, 232
Google, 218
Goose Island, 195
Great Depression, 83, 156
Great Fraser Midden, 144
Great Northern Expedition (Second Kamchatka Expedition), 3–4, 8, 10–11, 13–14, 16–38, 224
group theory, 95–98
Groves, Leslie, 108, 109
Gulf of Alaska, 35, 176, 191, 192
Guzmán, Nuño de, 45

Haefeli, Verena (née Huber), *see* Huber-Dyson, Verena
Hakai Passage, 195
Hale, Thomas, 15
Hall, John, 149–56, 158, 161
Halleck, Henry W., 180
Halliday, James "Scotty," 153
Halm, Katarina (née Haefeli), 97–99, 131
Handcock, John, 151
Hanson Island, 132–34, 186, 192
Hardy, G. H., 88
Hare, W. H., 241
Harper, Henry John, 203
Harper's Magazine, 83

Hartley, Ralph, 106
Harvard *Crimson*, 210
Has-yanoch, Tom, 150
Hawaii, 145–46, 170, 195
Heath, John, 184
heliography, 8–10, 39–40, 42, 61–63, 71, 221, 233
Heyerdahl, Jens Conrad, 170
Heyerdahl, Liv, 170
Heyerdahl, Thor, 168–70, 184, 194–96
Hilbert, David, 247
Hill-Tout, Charles, 144
Hiroshima, 109, 110, 112, 234
Hiscox, David, 158
Hogan, John Vincent Lawless, Jr., 80
Hooke, Robert, 228–30
Hoonah, 35, 190
Hoover, J. Edgar, 108
Horn, Tom, 53
Hornaday, William Temple, 239
Horn Cloud, Joseph, 243
Hotchkiss guns, 242
Howe, Bob, 191
Hrdlička, Aleš, 181
Huber, Charles, 94
Huber-Dyson, Verena, 94–99, 129; Freeman and, 98–99, 119
Hudson's Bay, 145
Hudson's Bay Company, 141, 148, 175, 179
humpback whales, 190
Huxley, Thomas, 203, 206
hydrodynamics, 92, 193
hydrogen, 85–86, 89–90
hydrogen bomb, 92, 93, 102, 109, 110, 113, 215, 224, 226, 232, 234, 245

I Ching, 223
Icy Strait, 189–91
identity tags: for Apaches, 10, 52, 54, 232–33; biometric, 234
India, 62
Indian Arm, 142–44, 148, 156
Indian River, 148, 156
Indian Wars, 40, 60, 62, 130, 180
infinities, 87, 99, 252, 253; in continuum hypothesis, 10, 246–47, 249
information, 106, 134, 249, 250; entropy and, 105–106

INDEX

Inside Passage, 137, 187
Institute for Advanced Study (IAS), 83–84,
  91–94, 97–101, 216, 224–25, 247; Electronic
  Computer Project, 92, 93, 101, 102, 212,
  215–17, 224–27
integrated circuits, 213
intelligence, 8, 207, 208, 245, 251, 253;
  artificial (machine), *see* artificial intelligence
intercontinental ballistic missiles, 105
International Committee of the Red Cross
  (ICRC), 95, 97
internet, 210, 214, 218, 232
iPhone, 215, 234
Ireland, 241
Irkutsk, 31
Ivy King test, 113
Izmailov, Gerasim Grigor'evich, 177–78

Jacobs, Mark, Jr., 37
Jameson, Annie, 75
Japan, 181; atomic bomb used against, 109,
  110, 112, 234
Jefferson, Thomas, 238
Jobs, Steve, 215
Jochelson, Waldemar, 181
Johnstone Strait, 132–33, 184, 186, 196–97
Jones, Henry Festing, 205
Joseph, Chief, 42
Jung, Imme, 119
Jupiter, 122
Jurasz, Charles, 190, 191

Kadin, Mikhail, 179
Kahn, Louis, 127
Kamchatka, Kamchadals, 3, 13, 16, 19–22,
  25–27, 29–32, 37, 171, 172–74, 177; First
  Kamchatka Expedition, 17; Second
  Kamchatka Expedition (Great Northern
  Expedition), 3–4, 8, 10–11, 13–14, 16–38,
  224
Karrer, Paul, 97
Kayak Island, 19
kayaks, 176, 181, 183, 222; author and, 129,
  131, 169, 182, 183, 185–86, 191–93, 195;
  turbulence and, 193–94
kayaks, Aleut (baidarkas), 10, 21, 24, 171,
  174–79, 180–82, 184, 187, 190–95; bow of,
  175, 181; life of, 194; stern of, 181–82

Kaywaykla, James, 52, 68
kelp highway hypothesis, 23, 143
Kemmer, Nicholas, 90
Kemp, George S., 76
Kennedy, John F., 125
Khitrov, Sofron, 19–22, 25, 26, 28
Khlebnikov, Kyrill, 176
Khmetevski, Vasili, 31
Khrushchev, Nikita, 128
Khvostov, Nikolai, 176
Kilburn, Tom, 217
killer whales (*Orcinus orca*), 31, 132–35
King, James, 145–46
Knight Inlet, 142
Kodiak, 172, 175–77, 179
Kolosh, 176
Kon-Tiki, 169–70, 184, 195
Kon-Tiki *Expedition, The* (Heyerdahl),
  168–69
Kroll, Norman, 86
Kuril Islands, 31
Kwakiutl (Kwakwaka'wakw), 132, 195

La Croyère, Louis Delisle de, 17, 19, 37
Lagunov, Ivan, 28
Lamarck, Jean-Baptiste, 206
Lamb, Willis, 85–86, 90
Lamb shift, 85–87, 89–90
Langsdorff, Georg Heinrich von, 178
languages, 133–35, 199, 211–12, 251
La Pérouse, Jean-François de Galaup, Comte
  de, 35, 191
lashing, sewing, and weaving, 9, 167–69, 183
Laughlin, William S., 23, 181
Lawton, Henry, 63–65
Lazukov, Aleksei, 22
learning, 245; deep, 232
Lefschetz, Solomon, 97
Leibniz, Gottfried Wilhelm, 3–10, 14–16, 39,
  94, 106, 165, 212, 223–24, 228, 230, 232,
  234, 246, 247, 251–53; *Discourse on the
  Natural Theology of the Chinese*, 223;
  Monadology, 223
Leni Lenapes, 100
Lepekhin, Thoma, 26
life: machines and, 208; tree of, 194
Life and Habit (Butler), 206
Littlewood, J. E., 88

Lituya Bay, 177, 191
Livermore National Laboratory, 110
Log Cabin Restaurant, 159, 160
logging, 137–40
logs, drifting, 137, 139–42
London *Reasoner*, 204
Looking Glass, Chief, 42
Loomis, Chuck, 115–16
Los Alamos National Laboratory, 84, 85, 89, 91, 108–15, 118, 126
Los Angeles Times, 58
Lowry, Malcolm, 157, 161
Lowry, Margerie, 157
Lozen, 60
Lummis, Charles, 58
Lyttelton-Christchurch telegraph line, 209–10

MacElmen, W. J., 153, 154
machines, 7, 209–10, 212–13, 217–19, 245, 252; evolution and, 208–209, 218; intelligence of, *see* artificial intelligence; language of, 211–12; life and, 208
Mack, Clayton, 170
Macy & Co., 83
magnetic fields, 169
Maksutov, Prince, 180
Malloff, Will, 192
Mance, Henry Christopher, 61–62
Mangas Coloradas, 49–50, 55, 66
Mangus, 66, 68–69
Manhattan Project, 84, 91, 108–109
Maori, 195
maps, mapping, and computation, 4, 6, 133–34, 248, 250–51
Maquinna, 146
Marconi, Giuseppe, 75
Marconi, Guglielmo, 75–77
Marconi Wireless, 75, 78
Marcos de Niza, Friar, 45
Mark, Carson, 112–14
Mars, 122, 125, 126, 131
Martin, Alan, 158
Martin, Carroll, 186–89
Mastrolilli, Paolo, 95
Mathematical and Numerical Integrator and Computer (MANIAC), 226–27
mathematics, 96, 212

Matthew, Patrick, 206
Matthews, J. S., 147
Matveev, Artamon, 14
Maus, Marion, 63
Maxwell, James Clerk, 73, 75, 78
Maxwell's demon, 106
Mayer, Harris, 123
McCabe and Mrs. Miller, 158
McCandless, H. W., 80
McGeer, Patrick, 134
McKinn, James (Santiago), 57–58
McKinn, Martin, 57
McLaughlin, James, 240
McReynolds, Andrew W., 111
Mead, Carver, 6
meaning, 212, 213, 250–51
"Mechanical Creation, The" (Butler), 204–205
memory, digital, 216–17, 222; human, 229–30
Mendoza, Antonio de, 45
Menzies, Archibald, 147
Mesopotamia (Butler homestead), 202–204
Mexicana, 144, 185
Mexican-American War, 48
Mexico, 42, 45–48, 56–57
Micrographia (Hooke), 229
microorganisms, 229
microprocessors, 6, 213, 223, 230, 249
microwaves, 85
Miles, Nelson Appleton, 40–43, 52, 56, 60–66, 69, 221, 233, 241
Miller, Jonathan, 152
mind, 4, 8, 133–35, 164, 207, 218, 229, 253
missiles, 105, 115, 116
MIT, 85
models, 193, 208, 250–51
molecules, 106, 208
Monadology (Leibniz), 223
Montreal Trust, 157–58
Moody, Sewell Prescott, 149
Moodyville, 149, 150
Mooney, James, 241–44
moon mission, 117, 125, 131
Moore's law, 250
Morgan, Thomas J., 241
Morse code, 39, 76–77
Mount Fairweather, 191–92, 196
Mount Waddington, 130

Muir, John, 130
mules, 53, 63
Müller, Gerhard Friedrich, 17, 19, 30, 31
Mulovskii, Grigorii Ivanovich, 173–74
music, 132–33
mutual assured destruction (MAD), 124
Myer, Albert J., 62

Nadezhda, 20
Nagai Island, 13
Nagasaki, 109, 110, 234
Naiche, 39, 40, 42–44, 52, 55–61, 64–67
Narváez expedition, 44
Naryshkin, Ivan, 14
NASA, 116, 123–26
National Bureau of Standards, 224
National Cybersecurity Initiative Data
    Center, 221–22, 228, 230, 233
National Harbours Board, 156, 161
National Park Service, 190
National Science Foundation, 124
National Security Agency, 10, 221
National Weather Bureau, 62
Na-tio-tish, 40, 41
Native American Church, 241
Native Americans, 237–39; Ghost Dance
    movement among, 10, 240–41, 243–44;
    Indian Wars, 40, 60, 62, 130, 180; origins
    of, 23, 43, 143; Wounded Knee massacre,
    241–43; *see also* Aleuts; Alutiiq; Apaches;
    Chugach; Kwakiutl; Sioux; Tlingits;
    Tsleil-Waututh; Zuni
Natural Theology (Paley), 203
nature, 6, 7, 52, 86, 96, 219, 222, 245, 252, 253;
    digital and analog computing in,
    248–49
Navy, U.S., 124
Nazi Germany, 95, 97, 105, 109, 170
NBC, 82
Neifert, William, 62
Nelson, Hugh, 149
nervous system, 8, 212, 245
Netscape, 211
networks, 6–7, 9, 39–40, 42, 61–63, 71, 164,
    211, 221–22, 233–34, 248–50
neurons, 7, 164, 248
Nevada Test Site, 118
New Army Pass, 130

Newton, Isaac, 7, 228
New Yorker, 93, 98
New Zealand, 10, 195, 200–201, 203, 209, 210
NeXT (computer), 215
Nez Percés, 42
Nikolski, 181, 182
Nimpkish Hotel, 186
Nipper (RCA Victor), 83
Noche-del-klinne, 54
noise, 217, 248
Nootka Sound, 146
Northwest Coast, 3, 8–11, 143, 145, 160, 163,
    165, 170–73, 195–96
nuclear energy, 107, 108, 110
nuclear physics, 91, 108
nuclear-propelled rockets, 115–26
nuclear reactor, 111–12
nuclear weapons, 9, 84, 92, 93, 102, 105,
    107–14, 118; global stockpile of, 124;
    hydrogen bombs, 92, 93, 102, 109,
    110, 113, 215, 224, 226, 232, 234, 245;
    intercontinental ballistic missiles, 105;
    Moscow-Washington hotline and, 128;
    mutual assured destruction and, 124;
    Szilard and, 9, 105, 107–10, 127–29; used
    against Japan, 109, 110, 112, 234;
    U.S.-Soviet agreements on, 123, 125
nucleotides, 6, 222, 248
numbers that mean things vs. numbers that do
    things, 225–27
Nyquist, Harry, 106

Okhotsk, 18, 174
Olden Farm, 91, 100, 101
Olden Manor, 91, 92
On the Origin of Species (Darwin), 200, 201,
    203–204, 206
Oppenheimer, Kitty, 91
Oppenheimer, Robert, 84, 91–93, 97, 98,
    102, 109–11, 113; Dyson's letter to,
    119–20
optical technologies, 10, 71, 233; heliographic,
    8–10, 39–40, 42, 61–63, 71, 221, 233
OrcaLab, 133
orcas (killer whales, *Orcinus orca*), 31,
    132–35
order codes, 227
organisms, 194, 298, 245, 249, 251

Origin of Mankind, The: Viewed from the Standpoint of Revelation and Research (Fleming), 73
Orion project, 9, 115–26, 131
ozone, 82

Paget, Percy Wright, 76
Paley, William, 203
Pascal, Blaise, 223
Paull, Andrew, 147
PC Forum, 213–15
Penn, William, 100
Peru, 170
Peter the Great, 3–5, 11, 14–18, 20, 224, 234
phones, 63, 214, 234; iPhone, 215, 234
physics, 86–91; nuclear, 91, 108; particle, 97–98; quantum, 73, 85, 90, 223
Piaget, Jean, 92
Pittendrigh, George, 154
plankton, 193
Pleistocene, 22, 167, 196, 240
Plenisner, Friedrich, 26, 27
plutonium, 114
Point Roberts, 147
Polynesia, 168, 170, 194, 195
Pomerene, James, 217, 225
Pomory, 172
Poole, Thomas, 153
Port Lyttelton, 202, 209
potlatch, 143
Powell, John Wesley, 241
Power, Thomas, 125
Predator (drone), 125
Preece, William Henry, 72
preindustrial epoch, 7
Prickett, Don, 118
Princeton, N.J., 82, 100
Princeton University, 82, 97, 130; Firestone Library, 129
Prince William Sound, 171, 172
Project Orion, 9, 115–26, 131
Project Plowshare, 110
promyshlenniki, 172, 173
protons, 88
Puget, Peter, 147, 148
Pump House Gang, The (Wolfe), 120
Pyrmont, 3, 15–16, 224

quantum mechanics, 73, 85, 90, 223
Quantum Theory of Fields (Wentzel), 90
Quatsino Sound, 182

radar, 81, 85, 250
radio, 75–83
railroads, 237, 238, 245
Randolph, G. W., 49
RCA, 82–83, 102, 224
RCA Victor, 82–83
refrigerator patent, 106, 107
RELease 1.0, 213
Resolution, 145, 177
Retherford, Robert C., 85
rifles, 51, 154
Rockefeller, John D., 156
Rockefeller Foundation, 91
rockets, nuclear-propelled, 115–26
Roman Emperor, 201
Roosevelt, Franklin D., 108
Roosevelt, Theodore, 56
Rosen, Benjamin, 213
Rosenberg, Jack, 224
Rosen Electronics Letter, 213
Royal Air Force (RAF), 88–89
Royal Institution of Great Britain, 74
Royal Navy, 124
Royal Society of London, 228
Russia, Russians, 3–5, 9–10, 15, 39, 148, 171–80, 190
Russian Academy of Sciences, 4, 16
Russian-American Company, 175–76, 178–80
Russian Naval Statute of 1720, 14
Russian Orthodox Church, 178
Rutherford, Ernest, 107
Ryffel, Berthy, 97

SABRE, 250
SAGE (Semi-Automatic Ground Environment), 250
Sv. (Saint) Apostol Peter, 171–72
St. Paul, 19, 32–37
St. Peter, 19–21, 24–25, 27, 29, 30, 32, 171
St. Petersburg, 15, 17, 31, 32, 173
Salish Sea, 146, 184
Salk Institute, 127
San Carlos reservation, 51–52, 54, 56–57, 66, 67

Sarnoff, David, 83
Sarychev, Gavriil A., 174, 177
satellites, 116; Sputnik, 114, 116, 117, 125
Saturday Review, 206–207
Saturn, 9, 122, 125, 126, 131
sawmills, 138, 140
Scalettar, Richard, 86
Scharff, Morris, 110
Schofield, John, 233
Schonchin, Peter, 240
Schwatka, Frederick, 61
Schwinger, Julian, 86
Science, 126
Scott Islands, 195
Sculley, John, 215
SEAC, 224
sea cows (*Hydrodamalis gigas*), 20, 29–30, 32
sea level, 22–23, 143, 195
sea mammals, 23, 143, 172, 175, 179, 182, 193, 195, 196, 239–40
sea otters (*Enhydra lutris*), 26, 28–30, 32, 175–76, 179–81
search engines, 251
Second Kamchatka Expedition (Great Northern Expedition), 3–4, 8, 10–11, 13–14, 16–38, 224
self-replication, 7, 208–209, 227, 252
semiconductors, 72, 213, 227
sequences, 5–6, 222–23, 227–28, 248
Serving the Republic (Miles), 52
Seven Cities of Cíbola, 45
700 Science Experiments for Everyone, 169
Sewid, James, 132
sewing, lashing, and weaving, 9, 167–69, 183
Seymour Narrows, 185
Shannon, Claude, 106, 252
Shaw, George Bernard, 200
Sheep, Charley, 243, 244
Shelikhov, Grigorii, 175, 177
Shelley, Mary, 232
Shelter Island, 86
Sheridan, Philip H., 59, 60, 66–67, 232
Sherman, William Tecumseh, 60, 232, 237–39
shift registers, 223–24, 226, 228, 230
Shiprock, 130
Shirland, Edmond D., 49
shitiki, 172
Shumagin, Nikita, 13

Siamoc, 149, 150
Siberia, 3, 18, 27, 31, 172, 173
Sibley, H. H., 49
Sieber, Al, 53
Sierra Club, 129–31, 188–89, 191
signal amplification, 78, 80–81
Signal Corps, U.S., 62–63, 221
silicon, 81, 102, 215, 227
Sioux, 238, 240–43
Sitka, 176, 179, 180
Skana (killer whale), 132–33
skin boats, 10, 23–24, 170–71, 182, 194–96
Sia-holt, James George, 149, 150
Slava Rossii, 174
Slutz, Ralph, 224–25
Smith, Alfred, 150
Smithsonian Institution, 180, 181, 183, 239
smoke signals, 42, 47
snapping turtles (*Chelydra serpentina*), 101, 103
Snow, C. P., 89
social networks, 250
software, 213
Soviet Union, 110; Moscow-Washington hotline, 128; nuclear agreements with U.S., 123, 125; Sputnik satellites of, 114, 116, 117, 125; Szilard and, 128
space exploration, 9, 108, 115–26
Spaniards, 43–47, 51, 144–45, 148
Speiser, Andreas, 95
Spong, Paul, 132–34, 192
Sputnik, 114, 116, 117, 125
squirrels, flying, 163
Stanley Park, 147, 157
Starboard Light Lodge, 142, 144, 156–59, 192
Starship and the Canoe, The (Brower), 191
Steller, Georg Wilhelm, 13–14, 17, 19–22, 24–32, 171, 190, 191; death of, 32
Stimson, Henry L., 109
Stone Age, 167–68
stone points, 167–68
Stoney, George Johnstone, 74
Strait of Juan de Fuca, 137
Strategic Air Command, 125
Strauss, Lewis L., 91
Streveler, Greg, 190–91
Sturgeon, D. B., 50

INDEX

Suggestions for Science Teachers in the Devastated Countries, 169
Super Oralloy Bomb (SOB), 113
Surge Bay, 36
surveillance, 221–22, 234
Sutil, 144, 185
Sv. Apostol Peter, 171–72
Sweden, 174
Szilard, Leo, 9, 105–10, 127; death of, 127; Einstein and, 106, 107; nuclear weapons and, 9, 105, 107–10, 127–29; *Voice of the Dolphins, The*, 9, 127–29, 134–35
Szilard, Trude (Weiss), 127

Tall Bull, 244
Taylor, G. I., 87
Taylor, Theodore "Ted," 111–17, 119, 120–23, 125, 128
technology, epochs of, 7–8, 135, 252
Tecumseh, 237–38, 241
telegraph, 57, 61–62, 222; Lyttelton-Christchurch line, 209–10
television, 81–83
Teller, Edward, 105, 109–11, 234
Tenskwatawa, 237–38, 241
Terent'ev, Koz'ma, 179
Terminal Steamship Company, 155
Tesla, Nikola, 78
Theory of the Leisure Class, The (Veblen), 83
Thomas, Dick, 94
Thomson, Joseph John, 74
thorium, 108
time, 10; in analog universe, 227, 228; in digital universe, 227–28; perception of, and body size, 229–30; tree rings and, 164–65
Timmerman, Franz, 14–15
Tlingits, 34–38, 171, 176, 180, 190, 194
toggles (electronic), 225
Torch Bay, 191
touch screen, 216–17, 219
traffic, 250, 251
transistors, 9, 102, 213, 223
trees, growth rings in, 9, 164–65
Tribolet, Robert, 59
TRIGA (Training, Research, Isotopes, General Atomic) reactor, 111–12
Trinity College, Cambridge, 88, 89
Trinity (Alamogordo) test, 109, 118

triodes, 78, 102, 225
Tripp, Charles, 202–203
Tripp, Ellen, 202–203
Trips to Satellites of the Outer Planets (Dyson), 122
Triquet Island, 196
Tri Sviatitelia, 177
Truman, Harry S., 109, 110
Tsleil-Waututh, 9, 143, 149, 155, 211
Tukey, John, 94
Tum-Tumay-Whueton, 143, 149
turbulence, 193–94
Turing, Alan, 74, 212, 216, 247, 252
turtles, snapping, 101, 103
Twyford School, 88

Ulam, Stanislaw, 114–15, 117, 219, 226, 245
Umnak Island, 23, 181
Unalaska, 174, 177, 178
Unangan, 23, 24; *see also* Aleuts
Under the Volcano (Lowry), 157
United Nations, 112
United Nations Educational, Scientific, and Cultural Organization, 169
University of California, 129, 131
uranium, 108

vacuum tubes, 6–7, 9, 10, 71–74, 78–82, 101–103, 215, 218, 223–25, 227, 249
Valdés y Flores, Cayetano, 144, 148, 185
Vancouver, 141, 142, 150, 153, 155, 156, 159, 160, 164
Vancouver, George, 144–48, 157, 184–85
Vancouver Island, 148, 170, 174, 179, 182, 195, 197
Veblen, Elizabeth, 92
Veblen, Oswald, 83, 84, 92, 100
Veblen, Thorstein, 83
Veniaminov, Ivan, 178–79
Verne, Jules, 116, 117, 121
Victoria, 148, 179
Victorio, 55, 56
Victor Talking Machine Company, 83
Vietnam War, 125, 130
Voice of the Dolphins, The (Szilard), 9, 127–29, 134–35
von Braun, Wernher, 116, 125
von Kármán, Theodore, 105

von Neumann, John, 6, 105, 106, 212, 213, 215, 224, 226, 232, 234–35, 247, 251, 252

Wade, James F., 67
Walker, Joseph Reddeford, 49
Wall Street Journal, 213
Warner, C. C., 243
War of 1812, 237
War of the Worlds (Wells), 117
Waxell, Laurentz, 18, 26, 28, 30
Waxell, Sven, 18–20, 22, 24–23
Way of All Flesh, The (Butler), 200
weaving, lashing, and sewing, 9, 167–69, 183
Weisskopf, Victor, 86
Wells, H. G., 107–108, 116, 117
Wentzel, Gregor, 90
West, Joseph R., 49–50
whales, 245; humpback, 190; killer, 132–35
whirlpools and eddies, 185–86, 193–95
White, John Bazley, 73
White Mountain Apaches, 40, 54
Whole Earth Catalog, 213
Widgeon, 186–87, 189
Wiener, Norbert, 106, 225, 245
Wigner, Eugene, 105
Wigwam Inn, 156, 158
Wilberforce, Samuel, 203
Williams, F. C., 217

Williams, Joshua Strange, 202
Williams tube, 216–17
Wodziwob, 240
Woese, Carl, 194
Wolfe, Tom, 120
Wood, Leonard, 63–64
World Set Free, The (Wells), 107
World War I, 82, 88, 89, 107–108
World War II, 84, 85, 88, 89, 91, 95, 108–109, 112, 121, 124, 130, 170, 181, 224, 225, 245, 249
World Wide Web, 210, 211
Worley, Ian, 158
Wounded Knee massacre, 241–43
Wovoka (Jack Wilson), 10, 240, 241, 243–44, 253
Wrangell Island, 29
Wren, Christopher, 15
Wright, John T., 50

Yakobi Island, 33–35
Yakutat, 171, 176, 177
Yakutsk, 17, 18, 30, 31
Yeats, John Butler, 206
Yellow Bird, 242
York, Herbert F., 117–18
Yushin, Kharlam, 24–25

Zuni, 44–47

A NOTE ABOUT THE AUTHOR

George Dyson is an independent historian of technology whose subjects have included the development (and redevelopment) of the Aleut kayak (*Baidarka*, 1986), the evolution of artificial intelligence (*Darwin Among the Machines*, 1997), a path not taken into space (*Project Orion*, 2002), and the transition from numbers that mean things to numbers that do things in the aftermath of World War II (*Turing's Cathedral*, 2012). He divides his time between Bellingham, Washington, and Harbledown Island, British Columbia, and is a dual citizen of Canada and the United States.